Mineral Processing and Recycling Technologies for Sustainable Future

Mineral Processing and Recycling Technologies for Sustainable Future

Guest Editors

Zoran Štirbanović

Bojan A. Marinković

Zhiyong Gao

Zhitao Feng

MDPI

Basel • Beijing • Wuhan • Barcelona • Belgrade • Novi Sad • Cluj • Manchester

Guest Editors
Zoran Štirbanović
Technical Faculty in Bor
University of Belgrade
Bor
Serbia

Bojan A. Marinković
Department of Chemical and
Materials Engineering (Deqm)
Pontifícia Universidade Católica
Rio de Janeiro
Brazil

Zhiyong Gao
School of Minerals Processing
and Bioengineering
Central South University
Changsha
China

Zhitao Feng
Department of Chemistry
Colorado State University
Fort Collins, CO
USA

Editorial Office
MDPI AG
Grosspeteranlage 5
4052 Basel, Switzerland

This is a reprint of the Special Issue, published open access by the journal *Minerals* (ISSN 2075-163X), freely accessible at: https://www.mdpi.com/journal/minerals/special_issues/AYGH2WR225.

For citation purposes, cite each article independently as indicated on the article page online and as indicated below:

Lastname, A.A.; Lastname, B.B. Article Title. *Journal Name* **Year**, *Volume Number*, Page Range.

ISBN 978-3-7258-7615-0 (Hbk)
ISBN 978-3-7258-7616-7 (PDF)
https://doi.org/10.3390/books978-3-7258-7616-7

Contents

About the Editors

Zoran Štirbanović

Zoran Štirbanović is a Full Professor at the University of Belgrade, Technical Faculty in Bor, Department of Mineral and Recycling Technologies. His research interests include mineral processing, flotation, recycling technologies, and the beneficiation of primary and secondary raw materials. He has participated in national and international research projects related to sustainable resource utilization and critical raw materials. His work focuses on improving separation efficiency and developing environmentally responsible processing technologies. He serves as a Section Editor of the *Journal of Mining and Metallurgy*, Section A: Mining.

Bojan A. Marinković

Bojan A. Marinković is an Associate Professor at Pontifical Catholic University of Rio de Janeiro and Head of the Department of Chemical and Materials Engineering. His main research areas are ceramic materials with a focus on synthesis, characterization, properties, and applications of ceramic nanostructures for environmental remediation, direct air capture of CO_2 and other energy transition applications, development of zero thermal expansion ceramics for thermal shock resistance, and processing of nanomaterials directly from mineral sands. He has a special interest in X-ray powder diffraction, spectroscopy, thermal analysis, and electron microscopy techniques as tools for materials characterization. He coordinates or has coordinated many research and innovation projects financially supported by major Brazilian and international industrial partners and enterprise customers. He published more than 125 research papers (H = 32) and serves as Editor for *Minerals* and *High-Temperature Materials*.

Zhiyong Gao

Zhiyong Gao received his Ph.D degree in Minerals Processing Engineering from Central South University (CSU), China in 2013, and has been a professor and vice dean in the School of Minerals Processing and Bioengineering at CSU since 2019. His research interests embrace mineral crystal chemistry, flotation, grinding, waste treatment etc. He has published more than 90 papers with a h-index of 45 and 6500 citations (Google Scholar). He also held more than 30 authorized patents. He has received numerous awards and honors including the IJMMM Young Scientist Award (2024), the Baosteel Excellent Teacher Award (2024), the Chief Scientist of the National Key Research and Development Program (2023), the National non-ferrous Metal Outstanding Youth Science and Technology Award (2022), the World's Top 2% Scientists by Stanford University (since 2021), National Green Mine Youth Science and Technology Award (2020), the China-Australia Young Scientist Exchange Award (2020), MEI Young Person Award (2019), and the Young Elite Scientists Award by China Association for Science and Technology (2017). Zhiyong has contributed to the scientific community by serving, for example, on the editorial boards of several journals (including *Physicochemical Problems of Mineral Processing* since 2016, *Minerals* since 2017, *Minerals Engineering* since 2021) and committee memberships. In 2017, Zhiyong initiated and chaired the 1st China International Conference of Young Scholars in Mineral Processing which is now a well-known annual event in mineral processing community.

Zhitao Feng

Zhitao Feng a full professor at the School of Minerals Processing and Bioengineering, Central South University. He completed his PhD at the University of California, Davis, and later did his postdoctoral research at Colorado State University. Currently, his work focuses on integrating machine learning with quantum chemistry to drive the digital design of flotation reagents. He has published more than 30 papers, including 10 first author papers in journals such as *Nature Chemistry, Journal of American Chemical Society, Minerals Engineering*.

Preface

The Reprint "Mineral Processing and Recycling Technologies for Sustainable Future" brings together selected contributions that reflect current scientific and technological progress in mineral processing and resource recovery. The growing demand for critical raw materials, combined with increasing environmental constraints, requires innovative solutions for the sustainable extraction and recycling of valuable resources. This Reprint provides an overview of developments in comminution, separation processes, flotation, hydrometallurgy, and recycling technologies. Particular attention is given to improving process efficiency, reducing environmental footprint, and enhancing the recovery of valuable materials from both primary and secondary sources. The contributions demonstrate how scientific research and engineering innovation support the green and digital transitions through responsible resource management. This Reprint is addressed to researchers, graduate students, industry experts, and decision-makers engaged in mineral processing and circular economy strategies.

Zoran Štirbanović, Bojan A. Marinković, Zhiyong Gao, and Zhitao Feng
Guest Editors

minerals

MDPI

Review

Towards Circularity in Serbian Mining: Unlocking the Potential of Flotation Tailings and Fly Ash

Nela Vujović [1], Vesna Alivojvodić [2], Dragana Radovanović [3], Marija Štulović [3], Miroslav Sokić [1] and Filip Kokalj [4,*]

1 Institute for Technology of Nuclear and Other Mineral Raw Materials, Bulevar Franš d'Eperea 86, 11000 Belgrade, Serbia; n.petronijevic@itnms.ac.rs (N.V.); m.sokic@itnms.ac.rs (M.S.)
2 The Academy of Applied Studies Polytechnic, Katarine Ambrozić 3, 11000 Belgrade, Serbia; valivojvodic@politehnika.edu.rs
3 Innovation Center of the Faculty of Technology and Metallurgy in Belgrade, University of Belgrade, Karnegijeva 4, 11120 Belgrade, Serbia; divsic@tmf.bg.ac.rs (D.R.); mstulovic@tmf.bg.ac.rs (M.Š.)
4 Faculty of Mechanical Engineering, University of Maribor, Smetanova 17, 2000 Maribor, Slovenia
* Correspondence: filip.kokalj@um.si

Abstract: This paper examines sustainable industrial practices in Serbia, particularly in the mining and energy sector, focusing on the potential of flotation tailings and fly ash, as materials with the largest share in disposed waste in Serbia in 2023 (95%). It highlights the environmental challenges of mining waste and explores innovative approaches to waste management within the circular economy framework. The study analyzes the current state of mining waste in Serbia, particularly in copper mining regions in the east of the country. It discusses the potential for metal recovery from waste and its reuse in various industries. The research also investigates the use of fly ash from thermal power plants as a valuable resource in the construction industry and other sectors. The paper reviews existing initiatives and legislation in Serbia in order to promote sustainable mining practices and waste utilization. By presenting case studies and potential applications, the study demonstrates how implementing circular economy principles in the mining sector can contribute to environmental protection, resource conservation, and economic growth in Serbia. The comprehensive overview of the current state in Serbia provides a solid foundation for establishing a higher degree of circularity in the mining and energy sectors.

Keywords: mining; flotation tailings; fly ash; Serbia; circular economy

1. Introduction

Mining is a key factor in the economic development of a country. As a pillar of economic and social development, mining improves living standards but simultaneously causes environmental degradation and increased consumption and depletion of natural resources. According to the global report on resources for the year 2024, published by the UN [1], the exploitation of raw materials has tripled in the last 50 years. The report also predicts that this number will increase by 60% if urgent and coordinated actions are not taken.

Large amounts of mining waste generated during the mining process pose a significant environmental hazard. Mining activities generate over 100 billion tons of solid waste yearly globally, representing a significant environmental challenge requiring appropriate management strategies [2]. It is estimated that the annual global production of metal mining waste is about four billion tons [3]. At the same time, over 1.2 billion tons of residual waste has already been accumulated in the European Union (EU) [4].

Mining waste generated during the exploitation, extraction, and processing of mineral raw materials can be in liquid form, as wastewater or acid mine water, and solid waste as flotation tailing. Additional descriptive waste categories are given in Table 1.

Table 1. Mining and mineral processing waste and its characteristics [5].

Category	Description	Characteristics and Impacts
Overburden	Material removed to access mineral deposits.	- Consists of soil and rocks, often in large volumes. - Low immediate pollution levels can cause soil erosion and loss of vegetation. - It is used for reclamation, but improper management can disrupt ecosystems.
Waste rock	Material with low concentrations of minerals is not economically viable for extraction.	- When exposed to air and water, it often contains sulfide minerals, leading to acid mine drainage (AMD). - Heaps can destabilize landscapes, causing landslides. - Pollution of nearby water sources is common, affecting aquatic life and human consumption.
Tailings	Finely ground material is generated during the processing of mineral resources.	- Contains residual minerals and flotation chemicals like cyanide or sulfuric acid. - Highly prone to leaching, contaminating groundwater and surface water. - Tailings storage facility failures can result in catastrophic environmental disasters, spreading pollutants over vast areas. - Dam failure.
Mine water	Water generated during mining activities.	- Includes acid mine drainage (AMD) rich in heavy metals and sulfates. - It can affect agricultural land, rendering it infertile. - Long-term pollution persists in water bodies, harming biodiversity and reducing drinking water quality.
Sludge	Sludge produced from mine water treatment in active facilities.	- Dense mixture of solids and treatment chemicals. - It may contain toxic elements such as arsenic, lead, or cadmium, posing significant disposal challenges. - Requires safe containment to prevent leaching into soil and water.
Gaseous waste	Gaseous emissions from mining processes.	- Dust from blasting can cause respiratory illnesses like silicosis. - Emission of gases like sulfur oxides (SOx) and nitrogen oxides (NOx) contributes to acid rain and air pollution. - Flotation and smelting fumes can be toxic to both humans and wildlife.
Fly ash *	Ash is produced during coal combustion in thermal power plants.	- High alkalinity due to its chemical composition. - It contains heavy metals like mercury, chromium, and lead, making it hazardous if improperly handled. - Accumulation in dumpsites causes dust storms and long-term contamination of soil and groundwater.

* Although fly ash belongs to the energy sector, due to the similar problem of depositing large quantities as flotation tailings, its use found its place in this work.

Fly ash and flotation tailings have been deposited in large quantities globally, making them important secondary resources [2,6]. The biggest challenges today, which are related to these two raw waste materials, are their proper disposal and the remediation of the consequences of irregular disposal [7].

Using fly ash and flotation tailings as secondary raw materials has numerous benefits, such as reducing the cost of the economic tax for landfilling and maintenance. However, it can have also a negative impact on the environment and human health [7].

Besides reducing the negative impact of fly ash on the environment and human health, their use as secondary raw materials offers numerous advantages, including lower economic costs for disposal and maintenance as well as a reduction in the land area covered by fly ash waste.

Mining waste disposal in its original form destroys fertile soil, forest areas, and surface and groundwater flows [8,9]. Finely dispersed particles from mine tailings and wind-borne fly ash [10] are deposited on crops and households, threatening resources such as soil, water, air, and food production by accumulating toxic substances. Toxic elements also spread vertically, disrupting ecosystems over larger areas and negatively affecting plant and animal communities and their habitats. Also, another of the key challenges associated with flotation tailings is the possibility of a dam bursting, which, regardless of other types of mining waste, represents a significant environmental problem. A potential breach of the flotation tailings dam releases a high-energy flood wave, endangering ecosystems downstream, including human and animal populations [11]. Unfortunately, such incidents are numerous and often occur due to natural disasters or technical defects in the construction of dams [12,13].

The deposition of flotation tailings and fly ash, in addition to causing significant consequences if the deposition is inadequately implemented, also has economic consequences.

Global annual fly ash production varies by source, with estimates ranging from 400 million [14] to over 750 million tons [15]. These figures underscore the substantial amount of fly ash generated worldwide and reflect differences in various data collection methods and the inclusion of different countries. Analyses indicate that the global fly ash market is projected to grow at an annual rate of approximately 4.83% from 2025 to 2033, driven mainly by the construction industry's demand for fly ash in cement and concrete [16].

In addition to the breach of dams, the deposited sulfide flotation tailings and ore tailings lead to the formation of acid mine drainage (AMD), which in the world literature received the epithet "many-headed beast" due to the wide range of negative impacts they have on the environment [17]. In the case of flotation tailings that have carbonate minerals, they can have buffering capacity and slow down the AMD formation process [5]. The formation of AMD usually, but not exclusively, occurs from iron–sulfide aggregate rocks [18]. Sulfide minerals from the outcrop, primarily pyrite (FeS_2), create acidic mine waters when exposed to oxygen, water, and a series of chemical and biological processes [19]. AMD remediation remains expensive, costing globally approximately USD 1.5 billion annually. In comparison, global environmental liabilities exceed USD 100 billion [20], which creates an additional obligation to remediate the flotation tailings as soon as possible.

Environmental concerns are common in developed and developing countries, and although their specificities may differ, both types of economies face similar challenges on the technical, economic, social, and legislative levels [21]. To better understand and solve these problems, it is crucial to thoroughly study the available data and the existing literature and compare different management strategies.

This paper aims to provide a comprehensive overview of research on this topic, focusing on the specific situation in the Republic of Serbia, a European but non-EU country with a developing economy. The review highlights the considerable potential for sustainable mining practices in Serbia, with a particular focus on flotation tailings and fly ash, management strategies, and current and innovative approaches for mineral mining and processing waste management that can be integrated into the circular economy concept. Also, an integrated system is proposed to increase the economic value of mineral mining and processing waste while reducing its adverse environmental consequences. Lastly, the study presents an overview of regulations concerning the management of extractive waste in the European Union and Serbia, discussing its implementation as a significant step towards sustainable practices in the Serbian mining sector.

2. The Principles and Possibilities of Circular Economy Within Mining Sector

A circular economy (CE) organizes the economy in such a way that materials are used as long and efficiently as possible, the need for new resources is reduced, and the amount of waste and harmful environmental effects are kept to a minimum. The CE concept offers numerous advantages to the mineral mining and processing sector. It enables the exploitation of valuable metals and production of high-value materials for other industries while reducing waste and soil pollution. This approach eliminates the need to use natural areas as landfills for mining and energy sector wastes. To achieve these goals, it is necessary to develop technologies for the recycling and recovery of materials, invest in modern infrastructure, support sustainable markets, and encourage cooperation between industry sectors, scientific institutions, governments, local communities, and consumers. Recent research has highlighted the importance of understanding the drivers and barriers to adopting CE principles to achieve efficient resource utilization [22]. Implementing these principles is closely linked to establishing conditions for the effective implementation of circular strategies. Recently, ten strategies were established as part of the circular economy 9R framework, which includes Refuse, Rethink, Reduce, Reuse, Repair, Refurbish, Remanufacture, Repurpose, Recycle, and Recover, (R0–R9, respectively) [23]. These strategies can be utilized in various mining operations, helping to close material loops and maximize value creation in the sector. Still, this study will focus on three strategies: reduction, reuse, and recycling, as it will also examine the generation and potential uses of mine tailings and fly ash. Recognizing the CE is vital, particularly in the mining, energy, and construction sectors, which are the three most significant drivers of the industry and also the largest sources of waste that can be repurposed as secondary raw materials.

Reduce: Reducing waste in the mining industry starts with mine planning, maximizing the valorization of existing resources (using new methods of ore enrichment, which would reduce the amount of tailings) and alternative sources of metals before opening new ones. In addition to using modern technology, it is necessary to work on reducing the exploitation of raw materials and more efficient use of resources as well as reducing energy needs in mining through advanced machines that consume less energy. The application of modern technologies, ground-penetrating radar (GPR), drones, and artificial intelligence (AI) and the automation of mining processes can significantly reduce waste materials and pollution, especially in tailings generation and AMD formation. The management of AMD is of particular importance, given that it is one of the most challenging environmental problems related to mining activities. Contamination of land and water systems caused by AMD seriously threatens natural ecosystems and human health, making treating these waters a key step towards sustainable mining development [24]. To keep AMD under control, flotation tailings must be disposed of in a controlled manner. Apart from AMD, AI can enable the selection of a more precise location when locating ore.

On the other hand, caution is necessary when using AI in mining. AI can lower the environmental and social costs associated with mining by utilizing advanced mineral processing and waste management technologies. While AI enhances machine efficiency, it also comes with a notable environmental footprint, which includes gas emissions, and the material costs related to its development and training [25].

Reuse: The reuse of mining waste refers to strategies that allow mining waste to be used without further processing or recycling, with the aim of reducing the amount of waste and prolonging its use in various industrial, construction, or environmental projects. Old mine waste dumps represent a significant source of raw materials, with advantages such as easy availability and fine particles that do not require additional processing [26–28]. These materials can be transformed into valuable products such as building materials (which will be discussed in more detail in this paper), including cement, bricks, and ceramics, and then fertilizers or materials for the restoration of degraded mining areas, if not characterized as hazardous waste that require further processing. Also, they can be utilized as backfill for mine shafts and closed mines, but care must be taken to dispose of them as an environmentally and mechanically stable waste [29,30].

Flotation tailings and fly ash, although characterized as potentially significant resources for a sustainable future, are still waiting for a breakthrough in implementation on an industrial scale, despite the achieved technological progress in ore processing in recent decades [31].

Recycling: Recycling mining waste enables the recovery of valuable metals, reducing waste and minimizing environmental risks. For example, a significant portion of steel, copper, and aluminum today is recycled from secondary sources. With the development of advanced separation and extraction technologies, the economic viability of tailings [32–34] and fly ash [35–37], treatment is becoming increasingly justified, opening up opportunities for new business models and contributions to local economies.

Flotation tailings represent an economically profitable source of metals and can be considered as a secondary deposit [38]. Since it is unnecessary to exploit the ore or crush it additionally, the costs of tailing treatment are significantly lower [32,39,40]. Also, the treatment reduces people's and the environment's exposure to contaminants. In addition to metal recovery, tailings can also be used as supplementary material in the construction industry, for example, to fill mines or in other industries [32,41–43].

After recovering available amounts of copper, gold, and sulfur into a concentrate, these materials can be used in smelting as an energy source or to produce sulfuric acid [44]. At the same time, reprocessing helps minimize the negative environmental impacts associated with acid mining drainage [45].

In agriculture, processed and neutralized tailings can be used as a soil additive or fertilizer containing phosphorus and potassium. This is especially useful in degraded areas, improving soil fertility and increasing agricultural yields.

The energy sector also uses tailings. Carbon-rich materials can be processed into briquettes and pellets as alternative energy sources. In addition, tailings containing rare earth elements are used to produce batteries and components for wind turbines, thus supporting the development of renewable energy sources [32].

Tailings also play a significant role in environmental projects, such as stabilizing riverbanks and rehabilitating mining areas. Mixed with organic matter, they can be used for afforestation and biodiversity conservation, providing that they do not emit toxic elements and are environmentally stable [46].

Industrial tailing applications include producing glass, paint pigments, wastewater treatment, and metal extraction for electronics and automobiles. Its capacity to remove heavy metals from wastewater demonstrates its importance in environmental solutions [47].

New research ideas point to tailings' potential to produce nanomaterials for medical and electronic applications [48] and to promote high-performance concrete [49]. Such innovations confirm that tailings can greatly enhance sustainable technologies' development and conserve natural resources when adequately treated.

In addition to industry, flotation tailings can be used to make decorative objects, sculptures, and tiles, supporting local crafts and small businesses. This approach contributes to local economic development and promotes sustainability [50].

The application of flotation tailings and fly ash is presented in Tables 2 and 3.

Table 2. The applications of flotation tailings in the construction sector.

Material	Application
Mine tailings (various types)	Recycling and reuse in construction materials [51,52]
Hematite tailings	Production of eco-friendly construction bricks [34]
Coal mining and processing wastes	Brick production and fuel for burning [53,54]
Copper mine tailings	Brick production [55]
	Production of fine-grained concrete [43,56]
	Production of paving stones as aggregates [56]
Iron ore tailings	Recycled aggregate concrete with polypropylene fibers [56,57]
Low-sulfide base-metal tailings	Rendering and masonry mortars [56]

Table 3. Applications of fly ash [58].

Applications/Category	Descriptions	Advantages
Construction industry	- Light aggregate in brick and blocks [59] - Cement replacement in concrete (up to 50%) improves durability, strength, and machinability [60] - Stabilize soil and embankments on roads	- Reduction in CO_2 emissions - Reducing the exploitation of natural resources - Improving the soil's bearing capacity and building material durability
Soil reclamation and stabilization	- Soil stabilization for infrastructure projects on soft surfaces - Filling and stabilization of mining pits	- Reducing the risk of erosion - Enabling renewable land use for agricultural or other purposes
Industrial application	- Neutralization of acidic industrial waters - Stabilization of hazardous materials - Production of lightweight aggregates and insulation materials	- Reducing environmental risks - Efficient management of resources
Environmental and energy applications	- Wastewater treatment (adsorbent for heavy metals and pollutants) - Binding of CO_2 through carbonization - Stabilization of mining pits	- Reduction in greenhouse gas emissions - Improving wastewater quality - Reducing the impact of mining activities

All these applications of fly ash contribute to closing the waste loop and reducing dependence on the exploitation of natural resources. Through its application, industrial waste is diverted from landfills to valuable products, preserving nature and reducing costs in construction and industrial projects. In this way, fly ash becomes an important resource in achieving the goals of the circular economy, which combines environmental, economic, and social benefits to achieve sustainable development [61].

3. Circular Economy Within Mining Sector in Serbia

Mining waste constitutes the most significant portion in Serbia, accounting for nearly 95% of the country's total waste production in 2023 [62]. This category encompasses waste generated during exploration, extraction from mines or quarries, and physical and chemical treatment of ores.

Regarding total quantities of hazardous and non-hazardous waste, according to Eurostat data [63], in the category of mining and quarrying, from 2010 to 2022, Serbia consistently ranked among the top six countries in Europe for mining waste generation. The country's position has steadily risen over the years: 2010–2012: sixth place; 2016–2018: fifth place; 2020: fourth place; and 2022: first place. In 2022, Serbia produced an astounding 24,542 kg of mining waste per inhabitant, including hazardous and non-hazardous waste.

Serbia also ranks prominently in the production of hazardous mining waste. From 2010 to 2022, the country consistently placed first or second in Europe for hazardous mining waste generation. The amount of hazardous mining waste per capita in Serbia has increased significantly: 2010–2020: approximately, 2000 kg per capita; 2022: about 4500 kg per capita (which places Serbia just behind Finland, which generated 5000 kg of hazardous mining waste per inhabitant in 2022) [63].

These statistics highlight the urgent need for improved waste management practices and environmental policies in Serbia's mining sector.

According to data from the Republic Statistical Office [64], the graphs in Figure 1 illustrate the millions of tons of mining waste and waste generated by the energy sector each year. After mining waste, the second largest category in volume is waste produced from electricity, gas, and steam supply processes, with fly ash representing approximately 80% of this industrial waste category [65]. About seven million tons of ash and slag are annually produced in the thermal power plants located in Serbia. Although it is known that ash is a valuable resource that is used all over the world in the construction industry to obtain cement and concrete in a more sustainable manner, in Serbia, only 3% is used for reuse [66].

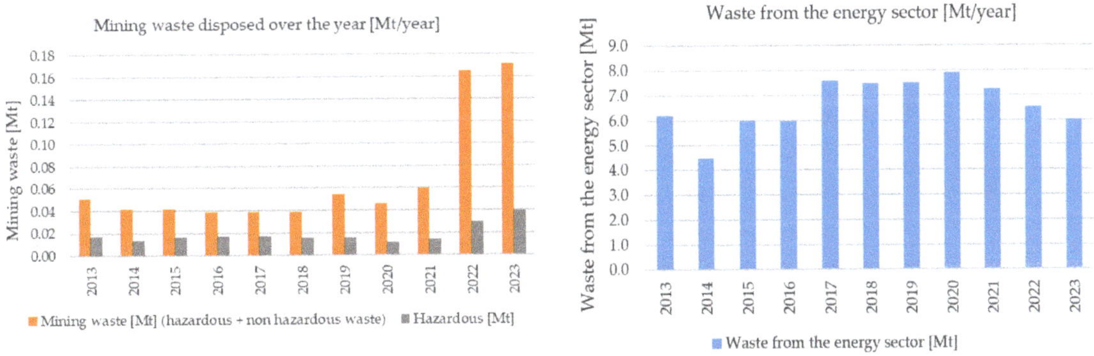

Figure 1. Generation of mining waste and waste from the energy sector over ten years in Serbia (generated upon data from [64]).

Figure 1 (left) shows the total amount of mining waste generated over the past 10 years, including both hazardous and non-hazardous waste. According to data from the Statistical Office of Serbia [64], the quantity of disposed mining waste has significantly increased in the last two reporting years compared to earlier years. However, due to the lack of more specific data, we can only speculate about the reasons for this increase. Figure 1 (right) displays the waste produced by the energy sector, with a significant portion coming from fly ash generated by burning coal in thermal power plants.

The annual average amount of mining waste deposited for the previous ten years is about 40 million tons of tailings and about 6.6 million tons of fly ash in the Republic of Serbia [64].

Flotation tailings, especially older tailings, often contain significant concentrations of metal compared to lower-grade ores. Namely, the concentration of copper in tailings can range between 0.2% and 0.4%, while in lower-quality ores, it ranges from 0.2% to 0.3% [67]. This can be explained by the application of outdated flotation technologies that were once in use, which did not enable a high degree of copper extraction. Thanks to technological advances, these methods today are much more effective. Considering the growing demand for copper and its current market prices, flotation tailings, as well as lower-grade ores, are increasingly viewed as a significant source for future exploitation [39,68].

In Serbia, the mining industry produces large amounts of waste, and one of the largest generators is primary copper production with eight active mines. In the Republic of Serbia, which has an area of 88,490 km^2, there are about 200 active mines and approximately 250 closed mines, according to the data of the Ministry of Environmental Protection of the Republic of Serbia [63]. These sites harbor vast amounts of deposited tailings, some of which have significant metal concentration data, making them potentially profitable for secondary mining. Figure 2 illustrates the geographical locations of the Bor, Majdanpek, Grot, Veliki Majdan, Rudnik, and Lece mines. The copper mines in Bor and Majdanpek belong to copper deposits, while Grot, Veliki Majdan, Rudnik, and Lece are Pb–Zn deposits. The following text describes their flotation tailings, located near the mine, along with a significant number of associated tailings [64].

Figure 2. A schematic representation of the locations of the biggest mining locations in Serbia.

The locations in Figure 2 present a dual challenge and opportunity. Tailings are an environmental problem, however, with appropriate analyses and technologies, they can be transformed into a resource of great value. This opens the possibility for a significant contribution to the sustainable development of the mining sector and the economically profitable use of secondary raw materials in Serbia.

Due to the significant copper production in Serbia, large amounts of waste material from primary copper processing were also created. The Bor and Majdanpek mines, which have been operating since 1903, have caused significant changes in the environment, serious geological disturbances, contamination of water resources, and disruption of biodiversity. The city of Bor is considered one of the most polluted sites in Serbia [69]. The Bor copper mine consists of three flotation tailings: the Old Flotation Tailings, the RTH Flotation Tailing Dump (hereafter, RTH Tailings), and the Veliki Krivelj Tailings. The Old Flotation Tailings and the RTH Tailings float ore obtained from underground mining, while the Veliki Krivelj Tailings are from surface mining.

The Old Flotation Tailings in Bor are located on the very border with the town of Bor. It consists of two smaller basins (Field 1 and Field 2) formed in the Bor River valley. The tailing dumps have an uncovered area of about 0.57 km^2, and a smaller part has been partially recultivated (about 0.07 km^2) [70]. It is estimated that 22.3 million tons of tailings are disposed of in this tailing dump, which potentially represents a commercially affordable source of gold, silver, and copper based on available data (Table 4) [71].

Table 4. Estimation of amounts of gold, silver, and copper affordable from disposed tailings [71].

Metal	[g/t *]	[t **]
Au	0.53	118
Ag	2.83	63.1
Cu	2300	51,290

* Share of metal per ton of disposed mining waste. ** Total mass of metal in the observed amount (22.3 million tons) of mining waste.

The active flotation tailing pit Veliki Krivelj has a total degraded area of 4.8 km^2 (of which 3.6 km^2 has been recultivated). It contains 190–195 million tons of tailings, with a copper content of about 0.10% [72].

The total copper reserves of the ore at Majdanpek were estimated at 800 Mt, containing 0.4–0.8% Cu and 0.25–1.0 g/t Au [73].

Two main flotation tailing ponds that operate within the Majdanpek copper mine are "Valja Fundata" and "Šaški Potok". By 2021, 350 million tons of tailings have been disposed of at 3 km^2, with a maximum layer thickness of 140 m, within the "Valja Fundata" tailing dump [71].

According to the data, 104 million m^3 of overburden and 65 million tons of gangue were deposited at location in Majdanpek [74]. In comparison, at the mine site in Bor, six million m^3 of overburden and two million t of gangue were deposited.

Flotation tailing pit "Grot", located in the south of the country, contains 550 million tons of tailings that include lead (Pb), zinc (Zn), silver (Ag), and cadmium (Cd); however, precise data regarding composition are not available. According to reports from the "Grot" mine for 2022 and 2023, 20 thousand tons of overburden and 5 million tons of gangue were also deposited at this location [72].

The Lece flotation tailing deposit, located in southern Serbia, belongs to the Pb–Zn ore deposits and contains 2–3 million tons of tailings, which contain gold (1.33 g/t), silver (3.64 g/t), and indium (In) [74].

The flotation tailing pit at Rudnik (Zlokućanski Potok) occupies an area of 400 thousand m^2. It holds 8.7 million tons of tailings, which have a high silver content of 1.1 g/t, along with aluminum at 1.34 g/t, bismuth (Bi), cadmium (Cd), and other elements that are significant for industrial use [72].

According to recent data, between 2022 and 2023, the Lecemine site received 80 thousand tons of overburden and 400 thousand tons of gangue. In comparison, the Rudnik mine site deposited 1.8 thousand m^3 of overburden and 500 thousand tons of gangue during the same period [74].

The Veliki Majdan tailing pond, located in the west of Serbia, spans an area of 8600 m^2 with an estimated total of over 1.9 million tons of tailings present. Chemical analyses indicate significant concentrations of heavy metals, including lead (Pb), zinc (Zn), copper (Cu), chromium (Cr), cobalt (Co), nickel (Ni), antimony (Sb), arsenic (As), mercury (Hg), as well as iron (Fe). Data from 2022 and 2023 show that 21 thousand tons of overburden and 32 thousand tons of gangue were deposited at the Veliki Majdan mine site [75].

Numerous scientific papers have investigated the possibility of valorization of flotation tailings from Serbia. Antonijević et al. [76] investigated the Cu leaching from the flotation tailings of the Bor copper mine in a glass reactor, adding iron chloride to a 0.1 mol/dm^3 solution of sulfuric acid as an oxidizer. The ratio of solid-to-liquid (S/L) phase was 1:5. After 30 h of leaching, a copper yield of 88% was achieved.

Stevanović et al. [77] carried out leaching on the same flotation tailing samples only in columns using sulfuric acid at a concentration of 0.1 mol/dm^3, with or without oxidizing agents in an S/L ratio of 1:1. The highest achieved copper yield was 89.87% during acid leaching in the column for a sample taken at a depth of 10 m without the addition of an oxidizing agent.

Stanković et al. [78] used an aqueous solution rich in acidophilic bacteria and iron obtained from the highly acidic lake "Robule" near Bor to leachate copper from the Bor flotation tailings in a rotary mixer. The average copper leaching rate was approximately 80%.

Sokić et al. [79] conducted experiments on leaching copper from oxide–sulfide ore from the Cerovo copper mine, which is also part of the Bor basin. The obtained degree of copper leaching was 73.8% after only three hours of leaching with a sulfuric acid solution with a concentration of 0.3 mol/dm^3. Stevanović et al. [80] tested the leaching of copper from the overburden of the Cerovo mine, which mainly contains copper oxide minerals, followed by a solvent extraction process. The total copper utilization throughout the process was 70%, confirming that this process can be applied in industrial conditions with positive economic results.

Recently, Stanković et al. [81] leached copper from the Bor flotation tailings with a solution of sulfuric acid (0.2 mol/dm^3). The solid residue after the leaching was further treated by the flotation concentration process. The average degree of copper removal in the leaching stage was 70%, while after re-flotation of the residue and its leaching, this degree increased to 83–85%. At the end of the process, the concentration of total copper was reduced three times compared to the content in the initial sample of Bor tailings before the leaching process [81].

These data indicate a significant potential for the valorization of metals from secondary sources in Serbia, many of which are on the list of critical raw materials of the EU [82].

In general, the basic concepts for disposed mining waste, flotation tailings, and fly ash utilization, following the circular economy principles, are shown in Figure 3. This concept can be divided into three main steps of material processing and residue extraction:

1. The primary step (primary extraction) involves
   (a) The usage of coal through combustion with residual fly ash obtained. Note: The fly ash, adequate for use in cement and/or concrete production, can only be

produced at high combustion temperatures or the combustion of coal results in fluidized ash with an irregular particle structure and unstable properties (often inadequate for mentioned applications) [83].

(b) The processing of ore to obtain concentrate and flotation tailings.

2. The secondary step focuses on revaluing metals from (a) fly ash and (b) flotation tailings, providing critical raw materials (CRMs) but with residual waste. Serbia's ore deposits are recognized as a significant source for obtaining mineral raw materials, including CRM [72].
3. Tertiary step: In the context of a circular economy, this step refers to the potential for reducing, reusing, and/or recycling the remaining waste to obtain material for further use. This may include, for example, stabilization processes like desulphurization and the utilization of leftover material within the construction sector.

Fly ash from thermal power plants can be effectively used in construction, applying the reuse and recycling principles of the circular principles. Its use as a binding material in interaction with cement increases the quality of concrete and construction while reducing cement consumption, which fits the reduction principles. This practice saves money and contributes to the circular economy by reducing waste and using resources more efficiently. This would reduce environmental impact and create new economic opportunities, exemplifying the circular economy approach [84].

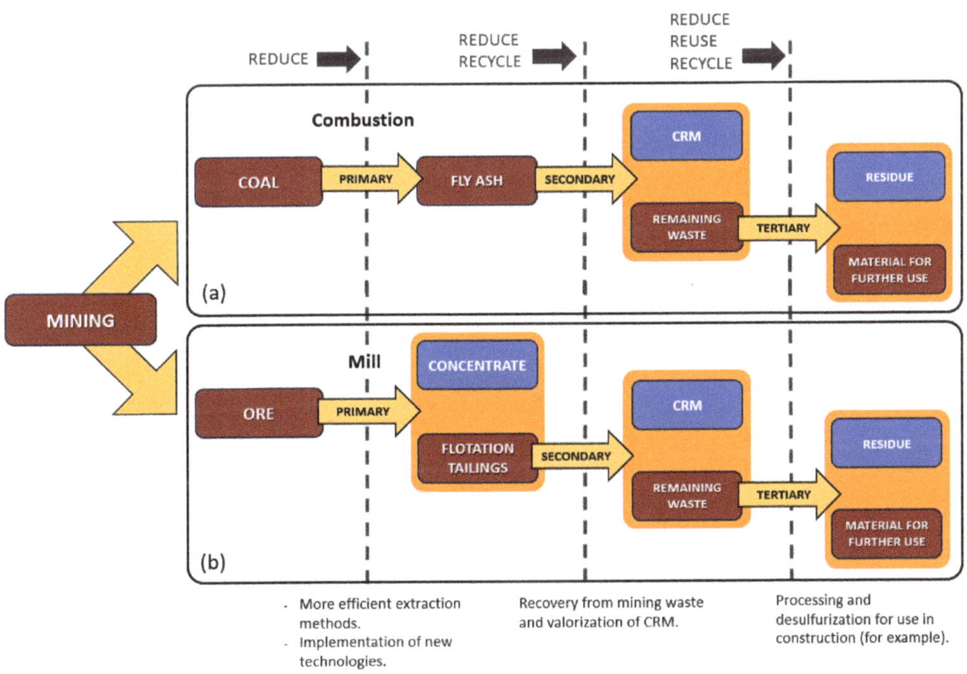

Figure 3. A flowchart of a three-stage separation process for (**a**) fly ash and (**b**) flotation tailings aimed at obtaining material for further use from leftovers.

Aside from illustrating potential uses for fly ash, Figure 3 also provides a comparative overview of mining tailing separation, which is discussed in more detail in the following text.

3.1. Reuse Potential of Mining Waste and Fly Ash in Serbia

Approximately seven million tons of fly ash and slag are produced annually in thermal power plants in Serbia, of which only 3% is used in the cement industry [85].

The Environmental Protection Agency of the Republic of Serbia receives reports from companies about the waste they generate and how it is managed. In 2023, the Agency reported that Serbia produced 7.84 million tons of waste, including 90,000 tons of hazardous waste. Thermal power plants are among the most significant waste producers in the country. Their coal combustion waste, classified as "10 01—Waste from power plants and other combustion plants", makes up 75% of the total waste, amounting to 5.87 million tons [63].

While this amount is lower than in previous years, it remains an environmental challenge, with most of it classified as non-hazardous waste (5.47 million tons) and stored at plant locations [86]. However, there are opportunities for sustainable management, as fly ash from coal combustion can be used in construction and cement production, contributing to a CE [63].

The iron and steel industry also generates significant amounts of slag processing waste, partly disposed of and reused. The export of non-hazardous waste, such as iron-containing metals, is a significant waste management segment [63].

There are three cement plants in Serbia: Lafarge Serbia (Beočin), CRH Serbia (Popovac, Paraćin), and Titan Cement Plant Kosjerić (Kosjerić). Lafarge and CRH have permits for the co-incineration of hazardous and non-hazardous waste as an alternative fuel, while Titan is in the process of obtaining a permit. These plants use over 300 thousand tons of non-hazardous and hazardous waste annually as an alternative raw material, including fly ash and granular slag from blast furnaces. This contributes to more efficient waste management and reduces environmental impact [63].

Due to its positive characteristics, such as mineralogical composition and high pH value, fly ash from thermal power plants in Serbia can be used for the treatment of hazardous waste. In the work of Radovanović et al. (2016) [87], fly ash was used to stabilize sludge from wastewater treatment in a copper smelter. By optimizing the addition of ash and other agents, stabilization of over 99% of metals (Cu, Pb, Ni, and Zn) and over 90% of As was achieved. The resulting solidified material had the characteristics of a non-hazardous waste suitable for disposal. Štulović et al. (2024) [88] investigated the possibility of flotation tailing geopolymerization using fly ash as a sustainable treatment of this type of mining waste. The resulting fly ash-based geopolymers had excellent structural, thermal, and mechanical properties, suitable for further application as a building material [88]. Đolić et al. (2021) [89] presented the circular use of fly ash. In their work, fly ash was first impregnated with goethite and then used as an adsorbent to remove As from wastewater. Value was added to this used/spent adsorbent by applying it as an adhesive in building materials with advanced mechanical properties [89].

Also, flotation tailings and fly ash have been successfully used to remediate AMD [90]. Laboratory tests have shown that flotation tailings from copper mine Majdanpek and fly ash from coal-fired power plants in Serbia ("Nikola Tesla" and "Kostolac") can be used to neutralize the highly acidic water of Lake Robule, situated near Bor in Serbia. When flotation tailings were used for neutralization, more than 99% of aluminum, iron, and copper were successfully removed, along with 98% of lead and 92% of zinc and about 90% of zinc and lead for both types of fly ash. The contribution of the results obtained in this research is an explicit and unambiguous indication that deposited waste, flotation tailings, and fly ash can be used to treat other types of waste, such as AMD. In this way, the negative impact of waste on the environment is reduced. Also, the solid residue after the AMD treatment is enriched with metals and can be subject to further testing in order to extract valuable components. On the other hand, treated AMD has such a chemical composition that it is safe for discharge into the environment. This result has emerged as good practice in using already deposited tailings, a bright spot in mining after long-term mining activity in Bor. This research marks an important step towards implementing sustainable development principles, particularly in terms of minimizing waste, treating it, and finding ways to reuse it within the same industry [90,91].

3.2. Legislation in Serbia

Serbia has taken significant steps towards implementing the CE in the mining sector. The Ministry of Environmental Protection has established a Working Group for the CE, while the Chamber of Commerce has established the CE Sector.

A notable step forward is the development of the Mining Waste Cadaster, financed by the EU, whose goal is to improve waste management. The Mining Water Cadaster also deals with treating and reusing water generated in mining processes. Adopting the by-product regulations allows mining waste to be turned into valuable raw materials for various industries, further contributing to sustainability.

The Mining Waste Cadaster project in Serbia, funded by the EU, was initiated to map and manage mining waste across hundreds of locations. By 2020, 150 landfills containing a total of 24 million m^3 of waste had been identified, with detailed examinations conducted on 41 of these landfills, which account for 80% of the total waste. The investigations involved analyzing soil, groundwater, and surface water as well as assessing landslide risks.

However, by 2024, the web application [74] only provided basic information about these locations, and a unified database had not been established. Experts have highlighted the necessity of transparent management in this area; however, an additional issue is the confusion surrounding data. More detailed information, including data on hazardous waste and site remediation, was located on the ministry's server but is not accessible to the public through the application.

State auditors have pointed out the existence of incomplete and inaccurate waste data, and mining waste continues to be managed without the required permits. Experts emphasize the urgent need for transparent and comprehensive management to minimize environmental risks [74].

This part of the paper covers the entire framework for the management of mineral mining and processing waste with a review of legislators' efforts to improve environmental laws and reduce the risks associated with mining waste through awareness programs (Table 5). A key objective of these regulations is to protect the communities living in the vicinity of mines and preserve the environment while encouraging responsible and ethical business.

Table 5. Legislation in the service of waste reduction.

Law/Regulation	
SERBIA	**Description**
Environmental Protection Act [92]	Establishes an integrated system for environmental protection, sets obligations for entities, and outlines the environmental permitting process.
Waste Management Act [93]	Regulates waste management, including mining waste, focusing on reducing environmental impacts.
Water Act [94]	Governs the protection and use of water resources, including regulations on wastewater discharge from mining operations.
Air Protection Act [95]	Sets air quality standards and measures to reduce industrial pollution, including emissions from mining activities.

Table 5. *Cont.*

Law/Regulation	
SERBIA	**Description**
Mining and Geological Research Act [96]	Regulates mining and geological activities, emphasizing environmental protection during operations.
Various regulations and decrees (e.g., Decree on Waste Categories and Recycling Standards) [97]	Includes specific regulations, such as the Decree on Waste Categories, detailing recycling standards for mining waste.
The End-of-Waste Status [98,99]	Rulebook on the types of waste for which an application may be submitted, permitted procedures and treatment technologies for types of waste, and other special elements for determining the end-of-waste status ("Official Gazette of the Republic of Serbia", No. 19/2024 and 47/2024) [98]: This regulation prescribes in detail the types of waste for which an application for end-of-waste status may be submitted, the permitted procedures and treatment technologies, as well as other special elements relevant for the determination of end-of-waste status. Regulation on technical requirements and other special criteria for certain types of waste that cease to be waste ("Official Gazette of the Republic of Serbia", No. 78/2019) [99]: This regulation prescribes technical requirements and specific criteria for certain types of waste (such as glass, paper, textile, aggregate, and metal), which, after meeting certain conditions, cease to be classified as waste.
European Union	**Description**
Landfill Directive [100]	It aims to prevent or reduce as much as possible any negative impact of landfills on surface water, groundwater, soil, air, or human health by introducing stringent technical requirements.
Directive 2006/21/EC on the Management of Waste from Extractive Industries [101]	Provides a framework for the sustainable management of mining waste, including preventive measures to minimize environmental impacts.
Directive 2008/98/EC on Waste [102]	Defines general principles for waste management, focusing on prevention, reuse, and recycling within a circular economy.
Raw Materials Initiative (COM(2008)699) [103]	This study is an important step towards understanding the recycling potential of mining waste in the EU. Its results provide the basis for a more sustainable approach to the use of raw materials and enable the development of effective policies that support the circular economy and sustainable development
Directive 2010/75/EU on Industrial Emissions [104]	Covers industrial emissions, requiring the best available techniques (BATs) for environmental protection.
Water Framework Directive (2000/60/EC) [105]	Establishes a framework for water protection and sustainable use, including monitoring and management of water pollution.
Directive 2014/95/EU on Non-Financial Reporting [106]	Requires companies to disclose environmental and social information, promoting transparency and sustainability.
EN 12620:2002+A1:2008; Aggregates for Concrete [107]	The standard specifies the requirements for aggregates in concrete, including natural, recycled, and manufactured types, to ensure quality and performance.

Table 5. *Cont.*

Law/Regulation	
European Union	**Description**
European Green Deal [108]	The European Green Deal is an EU initiative to achieve climate neutrality by 2050 through sustainable resource use, biodiversity protection, pollution reduction, and a just transition across all sectors.
Circular Economy Action Plan [109]	Outlines EU's strategic approach to circular economy, focusing on reducing waste, increasing recycling, and fostering sustainable industries.
New Construction Products Regulation (CPR) [110]	Establishes harmonized rules for marketing construction products in the EU (replacing regulation (EU) No. 305/20111) in order to improve legal clarity, reduce administrative burdens, and support the EU's climate and circular economy goals while ensuring safety.

Table 5 provides an overview of the regulations on managing extractive waste in the EU and Serbia. The table classifies laws that provide guidelines for the reclamation of mining areas and water and air conservation, contributing to sustainable development and environmental protection.

The core principles of a CE are straightforward: preventing waste and pollution, extending the lifespan of materials and products through repair and reuse, and restoring natural ecosystems. To enhance the implementation of the CE in Serbia's mining sector, it is essential to further develop new technologies and practices for effective waste management. Increasing the efficiency of resource recovery processes, improving waste neutralization methods, and implementing innovative recycling strategies are key to long-term results. Decision makers should provide support through appropriate regulatory and financial frameworks that encourage the recycling and reuse of mining waste, thereby paving the way for more sustainable and environmentally responsible mining. Raising public and industry awareness of the benefits of the CE through educational campaigns and research support can accelerate the broader application of these principles.

For the successful implementation of the CE in the mining waste and fly ash management sector, it is necessary to establish a national register of these materials as an initial source of data for resource recovery projects. This register should contain detailed information on the quantities, compositions, and potential uses of tailings and ash at different locations. In parallel, it is necessary to intensify research and development of technologies for efficient extraction of valuable metals from tailings and for broader application of fly ash in construction and other industries. Strong cooperation between scientific research institutions, industries, and government authorities is crucial to ensure that innovative solutions can find their way to practical application within these sectors. Additionally, it is necessary to develop and implement stricter standards for the characterization and use of these materials, especially their impact on the environment and human health. Finally, it is necessary to create an encouraging economic environment through tax incentives or subsidies for companies that actively apply the principles of the circular economy in managing mining waste and fly ash.

3.3. European Practice and Serbian Perspective for Mining Waste Management

In the European Union countries, many efforts have been made in the mining sector to prevent, reduce, and minimize the negative impacts of the mineral extraction activities and flotation tailing disposal on the environment and nature. These efforts are summarized in the Best Available Techniques (BATs) Reference Document for the Management of Waste from Extractive Industries [111]. Duly tailing management requires a good knowledge of the waste and its behavior over time and depends on various factors such as the mineralogical, chemical, mechanical, and physical properties of the waste. To assess the potential impacts of deposited mining waste over time, the European Commission suggests the EN 15875 test [111], a static test for the determination of acid potential and neutralization of sulfidic waste, as well as the EN 12457 (1–4) [111] test for the classification of waste as inert, non-inert, non-hazardous, or hazardous. This test is also adopted in Serbian legislation as a test for the classification of waste intended for disposal at landfills of inert/non-hazardous/hazardous waste [97]. Mineralogy is mostly the main characteristic of the tailings that determines the method of its treatment, further processing, or disposal. In general, oxide, sulfide, silicate, and carbonate minerals can be distinguished in the waste, which could further undergo fundamental chemical changes (e.g., weathering of sulfides to oxides). It is also important to have a complete mechanical and physical characterization of the tailings including shear strength, particle size distribution (which influences shear strength), density, plasticity, moisture content, permeability (hydraulic conductivity), etc. The main techniques for treating mining waste are (a) pre-sorting and selective handling that separate potentially valuable materials from the waste stream before the waste itself is sent for treatment and/or disposal; (b) physical treatment aiming at phase separation (evaporation, gravity separation, centrifugation, filtration); and (c) chemical treatment with the aim to reduce the negative environmental impacts of the waste (precipitation and separation, neutralization, stabilization/solidification, coagulation and flocculation, desulfurization, removing or destroying cyanides, etc.).

In the management of mining waste, the following waste management hierarchy of actions should be followed: first, the prevention of solid waste generation, then reduction in waste generation, and later, the reuse and recycling of mining waste. In order to prevent the generation of mining waste, techniques, such as pre-sorting and selective handling of materials as by-products/products, return to excavation cavities, and use of mining material for internal or external purposes (for site rehabilitation or for sale on the market as construction products), are used. In order to reduce the generation of mining waste, it is the practice to send the waste off-site for appropriate treatment and/or disposal or to apply appropriate separation of hazardous, non-hazardous, and inert waste with selective handling of each of them. Reprocessing flotation tailings to obtain valuable components is only applicable if it is technically and economically feasible and environmentally acceptable, which depends on the characteristics and quantity of the waste.

In the Republic of Serbia, the implementation of such actions is only in its infancy. In the Bor and Majdanpek districts, with a long history of copper ore extraction, a new stage in the reclamation of tailings on the territory of copper ore mines has only just begun. Grassing and afforestation of the dams of the Veliki Krivelj flotation tailing pond have been carried out, thereby preserving and restoring habitats, improving soil properties, reducing erosion, and enriching the soil with organic matter, as well as protecting other natural resources, such as air and water.

The reuse of fly ash produced by burning coal as a substitute for natural minerals or industrial products reduces the exploitation of raw materials, reducing energy consumption and CO_2 emissions. For example, one ton of fly ash replacing cement can reduce CO_2 emissions by approximately 600 kg [83]. In Serbia, the reuse of fly ash is quite low compared to the EU, where about 70% of fly ash is used in the production of cement, concrete, and concrete products [83]. In order to be used in construction, fly ash must have certain physical and chemical characteristics. For example, for the use in concrete according to EN 206 (European standard for concrete), the following characteristics are important: Loss on

Ignition (LOI) and the presence of chlorides and free calcium oxide, while for use in cement, the physical, chemical, and mechanical parameters of cement are determined by the EN 197-1 standard and the values for LOI, sulfates, and chlorides are monitored [83]. In order to obtain fly ash of favorable quality for further use and to reduce the proportion of unburnt carbon, the combustion process has to be optimized to achieve complete combustion conditions [83]. In addition to this combustion optimization, desulphurization methods are also applied to obtain the highest quality of fly ash. Apart from European standard EN 197-1, American standard ASTM C618 for the classification of fly ash in construction is also in force [112]. These combustion optimizations ensure that fly ash of satisfactory quality is obtained, and its further use is made possible. From an economic point of view, as of 2023, the cost of fly ash-blended cement, a traditional and commercially used clinker substitute, ranged from 25 to 30 USD/ton of CO_2. In comparison, natural pozzolana, a sustainable and emerging alternative to clinker, costs 70 to 75 USD/ton of CO_2 [113], making it an attractive option for construction projects. In Serbia, three out of eight thermal power plants have put standardized fly ash on the market, while deposed ash is sold at significantly lower prices [114]. Considering that some EU countries, like Germany and Poland, have reduced coal production while being technologically advanced and already utilizing a large share of fly ash for economic purposes, Serbia could, in the future, benefit by increased utilization at home or by positioning itself as a potential exporter of fly ash. To seize this opportunity, Serbia needs to adjust its legislation policies accordingly.

There are no defined best-available techniques for treating fly ash that has unsuitable characteristics for further use in the construction industry, apart from its safe disposal. One potential solution is fly ash inertization, a process that reduces its toxicity. This allows for its disposal in landfills at lower costs while minimizing environmental risks [115].

While recycling mining waste into construction materials brings environmental and economic benefits, it needs further investigation due to the presence of heavy metals such as mercury, arsenic, and lead [116]. These harmful substances can be released into the air, water, and soil, endangering health through dust inhalation, contact, or contaminated water, especially in residential and industrial buildings where children are particularly vulnerable. Therefore, strict regulations, quality monitoring, and the application of stabilization/immobilization technologies are necessary to reduce risks and ensure the safe use of these materials. Additionally, utilizing fly ash and tailings in construction supports a circular economy model, where waste from one industry is repurposed as a resource for another.

Meanwhile, the construction industry is witnessing significant changes and new challenges. With the new ISO 5900 series of circular economy standards (ISO 59004, ISO 59010, and ISO 59020, introduced in 2024 [117]), significant progress has been made towards adopting circular economy practices globally, including in the construction sector. The sector's objectives have shifted to using more sustainable materials such as mine tailings and fly ash. Still, companies have been put in a position to avoid accusations of "greenwashing" (a term reflecting misleading investments or false prerogatives about the environmental benefits of products or services) to meet increasing investor expectations. The new EU Construction Products Regulation [111], which entered into force in January 2025, further emphasizes the importance of sustainability and digitalization, promoting recycled materials and innovative techniques.

4. Conclusions

The Republic of Serbia is among the top six European countries for the production of both non-hazardous and hazardous mining waste. In terms of hazardous mining waste specifically, it consistently ranked first or second from 2010 to 2022. This paper indicates significant potential for sustainable mining practices in Serbia, focusing on flotation tailings and fly ash. These materials, often viewed as waste, can be converted into valuable resources within a CE framework.

Flotation tailings, especially from older deposits, have been found to contain high concentrations of valuable metals such as copper, gold, and silver. Advances in processing technologies now allow for the economic reprocessing of these tailings, which reduces environmental impact while recovering valuable resources. Case studies from various mining sites in Serbia—including Bor, Majdanpek, and Veliki Majdan—demonstrate the potential for metal recovery and environmental remediation.

Fly ash, produced in large quantities by thermal power plants in Serbia, presents another significant opportunity for sustainable resource utilization. Data could be more promising, with only about 3% of it utilized in the cement industry, indicating significant underuse. However, fly ash has potential applications in construction, agriculture, and environmental projects. Utilizing fly ash as a secondary raw material could significantly decrease waste accumulation and reduce the need for natural resource extraction.

Implementing CE principles in Serbia's mining sector faces challenges, including regulatory barriers and the need for technological investment. Recent initiatives, such as the establishment of the Working Group for Circular Economy and the development of the Mining Waste Cadaster, demonstrate a growing commitment to sustainable practices. However, the Mining Waste Cadaster continues to face ongoing issues with data accuracy and accessibility.

In European Union countries, there are accepted reference documents and standards that prescribe a waste management hierarchy, from the separation of hazardous and non-hazardous waste streams to recommendations for treatment and disposal (hazardous waste) and use (non-hazardous) on-site and off-site. To unlock the potential of mining waste and fly ash in Serbia, it is crucial to enhance the legislative framework, invest in research and development, and encourage collaboration between industry, academia, and government. Utilizing advanced technologies, mining waste can be converted into secondary raw materials for the cement and ceramic industries, fostering waste reduction, environmental protection, and new economic opportunities within a circular economy.

The transition to a CE within Serbia's mining sector needs numerous steps to be fully implemented. However, it represents a potent chance for developing different sectors, obtaining an upgraded level of environmental protection, helping resource conservation, and contributing to economic growth. Serbia can pave the way for a more sustainable and prosperous future in mining and related industries by harnessing the potential of flotation tailings and fly ash.

Author Contributions: Conceptualization, methodology, and investigation, N.V. and V.A.; writing—original draft preparation, N.V., V.A. and D.R.; data curation M.Š.; writing—review and editing, N.V., V.A., F.K., D.R., M.Š. and M.S.; supervision, F.K. and M.S.; funding acquisition, D.R., F.K. and M.S. All authors have read and agreed to the published version of the manuscript.

Funding: This work was financially supported by the Ministry of Science and Technological Development and Innovation of the Republic of Serbia (Contract No. 451-03-136/2025-03/200023 and 451-03-136/2025-03/200287) and by the Slovenian Research and Innovation Agency (Research Core Funding No. P2-0196).

Conflicts of Interest: The authors declare no conflicts of interest. The funders had no role in the design of the study; in the collection, analyses, or interpretation of data; in the writing of the manuscript; or in the decision to publish the results.

References

1. Dale, G. Extraction of Raw Materials Could Rise 60% by 2060—And Making Mining 'Greener' Won't Stop the Damage. The Conversation 2024. Available online: https://theconversation.com/extraction-of-raw-materials-could-rise-60-by-2060-and-making-mining-greener-wont-stop-the-damage-222818 (accessed on 16 February 2024).
2. Lèbre, É.; Corder, G.D.; Golev, A. Sustainable Practices in the Management of Mining Waste: A Focus on the Mineral Resource. *Miner. Eng.* **2017**, *107*, 34–42. [CrossRef]
3. Haas, W.; Krausmann, F.; Wiedenhofer, D.; Heinz, M. How Circular is the Global Economy? An Assessment of Material Flows, Waste Production, and Recycling in the European Union and the World in 2005. *J. Ind. Ecol.* **2015**, *19*, 765–777. [CrossRef]
4. Wang, C.; Harbottle, D.; Liu, Q.; Xu, Z. Current State of Fine Mineral Tailings Treatment: A Critical Review on Theory and Practice. *Miner. Eng.* **2014**, *58*, 113–131. [CrossRef]
5. Dimitrijević, M.D.; Milić, S.M. *Sulfidni Rudarski Otpad: Karakteristike, Uticaj na Životnu Sredinu i Tretman*; Univerzitet u Beogradu, Tehnički Fakultet: Bor, Serbia, 2017; ISBN 978-86-6305-063-1.
6. Abaka-Wood, G.B.; Addai-Mensah, J.; Skinner, W. The Concentration of Rare Earth Elements from Coal Fly Ash. *J. S. Afr. Inst. Min. Metall.* **2022**, *122*, 7–15. [CrossRef]
7. Karanac, M. The Utilization of Fly Ash Modified by Calcium Hydroxide and Ferric Oxide for the Removal of Heavy Metal Ions from Water. Ph.D. Dissertation, University of Belgrade, Faculty of Technology and Metallurgy, Belgrade, Serbia, 2018.
8. Hakkou, R.; Benzaazou, M.; Bussiere, B. Valorization of Phosphate Waste Rocks and Sludge from the Moroccan Phosphate Mines: Challenges and Perspectives. *Procedia Eng.* **2016**, *138*, 110–118. [CrossRef]
9. Marković, R.; Marjanović, V.M.; Stevanović, Z.; Gardić, V.; Petrović, J.; Kovačević, R.; Štirbanović, Z.; Friedrich, B. Importance of Changes in the Copper Production Process through Mining and Metallurgical Activities on the Surface Water Quality in the Bor Area, Serbia. *Metals* **2024**, *14*, 649. [CrossRef]
10. Pierwoła, J.; Ciesielczuk, J.; Misz-Kennan, M.; Fabiańska, M.J.; Bielińska, A.; Kruszewski, Ł. Structure and thermal history of the Wełnowiec dump, Poland: A municipal dump rehabilitated with coal waste. *Int. J. Coal Geol.* **2018**, *197*, 1–19. [CrossRef]
11. Trifunović, V.; Avramović, L.; Božić, D.; Jonović, M.; Šabaz, D.; Bugarin, D. Flotation Tailings from Cu-Au Mining (Bor, Serbia) as a Potential Secondary Raw Material for Valuable Metals Recovery. *Minerals* **2024**, *14*, 905. [CrossRef]
12. Ke, X.; Zhou, X.; Wang, X.; Wang, T.; Hou, H.; Zhou, M. Effect of Tailings Fineness on the Pore Structure Development of Cemented Paste Backfill. *Constr. Build. Mater.* **2016**, *126*, 345–350. [CrossRef]
13. Tayebi-Khorami, M.; Edraki, M.; Corder, G.; Golev, A. Re-Thinking Mining Waste through an Integrative Approach Led by Circular Economy Aspirations. *Minerals* **2019**, *9*, 286. [CrossRef]
14. Trace.org.au. Exploring the Potential of Fly Ash. Available online: https://trace.org.au/exploring-the-potential-of-fly-ash/ (accessed on 23 January 2025).
15. Toniolo, N.; Boccaccini, A.R. Fly ash-based geopolymers containing added silicate waste: A review. *Ceram. Int.* **2017**, *43*, 14545–14551. [CrossRef]
16. IMARC Group. Fly Ash Market Size, Share, Trends and Forecast to 2033. Available online: https://www.imarcgroup.com/fly-ash-market (accessed on 23 January 2025).
17. RoyChowdhury, A.; Sarkar, D.; Datta, R. Remediation of Acid Mine Drainage-Impacted Water. *Curr. Pollut. Rep.* **2015**, *1*, 131–141. [CrossRef]
18. Dimitrijević, M.D.; Alagić, S.Č. Passive Treatment of Acid Mine Drainage. *Bakar* **2012**, *37*, 57–68.
19. Petronijević, N.S. Analysis of the Mechanism of Acid Mine Drainage Neutralization Using Flotation Tailings and Fly Ash. Ph.D. Dissertation, University of Belgrade, Faculty of Technology and Metallurgy, Belgrade, Serbia, 2022.
20. Manchisi, J.; Ndlovu, S. Prediction of Acid Mine Drainage Formation. In *Acid Mine Drainage*; CRC Press: Boca Raton, FL, USA, 2021; pp. 31–56. [CrossRef]
21. Saleem, S.; Hu, G.; Li, J.; Hewage, K.; Sadiq, R. Evaluation of Offshore Oil Spill Response Waste Management Strategies: A Lifecycle Assessment-Based Framework. *J. Hazard. Mater.* **2022**, *432*, 128659. [CrossRef]
22. Alivojvodic, V.; Kokalj, F. Drivers and Barriers for the Adoption of Circular Economy Principles towards Efficient Resource Utilisation. *Sustainability* **2024**, *16*, 1317. [CrossRef]
23. Alivojvodic, V.; Kokalj, F. Redefining Waste: R-Strategies and Metrics as a Framework for Driving Progress of Circular Economy Performance. In *New Trends in Engineering Research 2024; in Lecture Notes in Networks and Systems*; Springer: Cham, Switzerland, 2024; Volume 1216. [CrossRef]
24. Petronijević, N.; Alivojvodić, V.; Sokić, M.; Marković, B.; Stanković, S.; Radovanović, D. Sustainable Mining Towards Accomplishing Circular Economy Principles. *J. Appl. Eng. Sci.* **2020**, *18*, 493–499. [CrossRef]
25. Van Wynsberghe, A. Sustainable AI: AI for sustainability and the sustainability of AI. *AI Ethics* **2021**, *1*, 213–218. [CrossRef]
26. Falagan, C.; Grail, B.M.; Johnson, D.B. Bioleaching of Metals from Mine Tailings. *Miner. Eng.* **2017**, *106*, 71–78. [CrossRef]
27. Shengo, L.M. Potentially Exploitable Reprocessing Routes for Recovering Copper and Cobalt Retained in Flotation Tailings. *J. Sustain. Met.* **2021**, *7*, 60–77. [CrossRef]

28. Lei, B.; Li, X.; Guo, Y.; Qu, F.; Zhao, C.; Tam, V.W.Y.; Wu, V.; Li, W. Recycling of Copper Tailing as Filler Material in Asphalt Paving Mastic: A Sustainable Solution for Mining Waste Recovery. *Case Stud. Constr.* **2024**, *20*, e03237. [CrossRef]
29. Ndlovu, S.; Simate, G.S.; Matinde, E. *Waste Production and Utilization in the Metal Extraction Industry*; CRC Press: Boca Raton, FL, USA, 2017. [CrossRef]
30. Connors, E. Coal-ash management by U.S. electric utilities: Overview and recent developments. *Util. Policy* **2015**, *34*, 30–33. [CrossRef]
31. Argane, R.; Benzaazoua, M.; Hakkou, R.; Bouamrane, A. A Comparative Study on the Practical Use of Low Sulfide Base-Metal Tailings as Aggregates for Rendering and Masonry Mortars. *J. Clean. Prod.* **2015**, *87*, 914–925. [CrossRef]
32. Abbadi, A.; Mucsi, G. A Review on Complex Utilization of Mine Tailings: Recovery of Rare Earth Elements and Residue Valorization. *J. Environ. Chem. Eng.* **2024**, *12*, 113118. [CrossRef]
33. Alcalde, J.; Kelm, U.; Vergara, D. Historical Assessment of Metal Recovery Potential from Old Mine Tailings: A Study Case for Porphyry Copper Tailings, Chile. *Miner. Eng.* **2018**, *127*, 334–338. [CrossRef]
34. Chen, Y.; Zhang, Y.; Chen, T.; Zhao, Y.; Bao, S. Preparation of Eco-Friendly Construction Bricks from Hematite Tailings. *Constr. Build. Mater.* **2011**, *25*, 2107–2111. [CrossRef]
35. Dardona, M.; Mohanty, S.K.; Allen, M.J.; Dittrich, T.M. From ash to oxides: Recovery of rare-earth elements as a step towards valorization of coal fly ash waste. *Sep. Purif. Technol.* **2023**, *314*, 123532. [CrossRef]
36. Thomas, B.S.; Bhattacharya, S. Quantification and recovery prospects of critical metals and rare earth elements in Victorian brown coal fly ash: A promising secondary source for critical metal extraction. *Hydrometallurgy* **2025**, *211*, 106428. [CrossRef]
37. Kokalj, F.; Alivojvodić, V.; Lešnik, L.; Petronijević, N.; Radovanović, D.; Samec, N. Enhancing Utilization of Municipal Solid Waste Bottom Ash by the Stabilization of Heavy Metals. *Sustainability* **2025**, *17*, 1078. [CrossRef]
38. Borbor, A.K.K.G.; Nwaila, G.; Manzi, M.; Ghorbani, Y.; Ndlovu, S.; Petersen, J. The Valorisation of Platinum Group Metals from Flotation Tailings: A Review of Challenges and Opportunities. *Miner. Eng.* **2023**, *201*, 108216. [CrossRef]
39. Adrianto, L.R.; Pfister, S. Prospective Environmental Assessment of Reprocessing and Valorization Alternatives for Sulfidic Copper Tailings. *Resour. Conserv. Recycl.* **2022**, *186*, 106567. [CrossRef]
40. Opara, C.B.; Blannin, R.; Ebert, D.; Frenzel, M.; Pollmann, K.; Kutschke, S. Bioleaching of Metal(loid)s from Sulfidic Mine Tailings and Waste Rock from the Neves Corvo Mine, Portugal, by an Acidophilic Consortium. *Miner. Eng.* **2022**, *188*, 107831. [CrossRef]
41. Gitari, M.W.; Akinyemi, S.A.; Thobakgale, R.; Ngoejana, P.C.; Ramugondo, L.; Matidza, M.; Mhlongo, S.E.; Dacosta, F.A.; Nemapate, N. Physicochemical and Mineralogical Characterization of Musina Mine Copper and New Union Gold Mine Tailings: Implications for Fabrication of Beneficial Geopolymeric Construction Materials. *J. Afr. Earth Sci.* **2018**, *137*, 218–228. [CrossRef]
42. Li, Z.; Guo, L.; Zhao, Y.; Peng, X.; Kyegyenbai, K. A Particle Size Distribution Model for Tailings in Mine Backfill. *Metals* **2022**, *12*, 594. [CrossRef]
43. Huynh, T.P.; Nguyen, T.M.; Lam, T.K. Utilization of High Volumes of Copper Mine Tailings in the Production of Fine-Grained Concrete. *Mater. Today* **2022**, *65*, 543–548. [CrossRef]
44. Özdemir, E.; Lang, A.; Saari, J.; Liipo, J. Recovery of Sulfur, Copper, and Gold by Reprocessing Old Flotation Tailings at Bor, Serbia. *Asp. Min. Miner. Sci.* **2024**, *12*, 000783.
45. Ejtemaei, M.; Nguyen, A. Characterisation of Sphalerite and Pyrite Surfaces Activated by Copper Sulphate. *Miner. Eng.* **2017**, *100*, 223–232. [CrossRef]
46. Lottermoser, B.G. Recycling, reuse and rehabilitation of mine wastes. *Elements* **2011**, *7*, 405–410. [CrossRef]
47. Ouyang, D.; Zhuo, Y.; Hu, L.; Zeng, Q.; Hu, Y.; He, Z. Research on the Adsorption Behavior of Heavy Metal Ions by Porous Material Prepared with Silicate Tailings. *Minerals* **2019**, *9*, 291. [CrossRef]
48. Pizarro Barraza, F.; Thiyagarajan, D.; Ramadoss, A.; Manikandan, V.S.; Dhanabalan, S.S.; Venegas Abarzúa, C.; Sotomayor Soloaga, P.; Campos Nazer, J.; Morel, M.J.; Thirumurugan, A. Unlocking the potential: Mining tailings as a source of sustainable nanomaterials. *Renew. Sustain. Energy Rev.* **2024**, *202*, 114665. [CrossRef]
49. Feng, W.; Yu, Z.; Bao, R.; Xiong, J.; Yan, K.; Liu, R.; Zhang, R.; Lu, X. From waste to resource: Advanced activation techniques for tailings in sustainable cement production. *J. Build Eng.* **2024**, *97*, 110780. [CrossRef]
50. Ikramova, Z.; Mukhamedzhanova, M.T.; Tukhtaeva, G.G. Tungsten–molybdenum ore flotation tailings for ceramic tile production. *Glass Ceram.* **2009**, *66*, 102–103. [CrossRef]
51. Araujo, F.S.M.; Taborda-Llano, I.; Nunes, E.B.; Santos, R.M. Recycling and Reuse of Mine Tailings: A Review of Advancements and Their Implications. *Geosciences* **2022**, *12*, 319. [CrossRef]
52. Flores-Badillo, J.; Salinas-Rodríguez, E.; Hernández-Ávila, J.; Vargas-Ramírez, M.; Flores-Hernández, J.A.; Rodríguez-Lugo, V.; Cerecedo-Sáenz, E. Design and Production of a New Construction Material (Bricks), Using Mining Tailings. *Int. J. Eng. Sci. Res. Technol.* **2016**, *6*, 225. [CrossRef]
53. Stolboushkin, A.Y.; Ivanov, A.I.; Fomina, O.A. Use of Coal-Mining and Processing Wastes in Production of Bricks and Fuel for Their Burning. *Procedia Eng.* **2016**, *150*, 1496–1502. [CrossRef]

54. Taha, Y.; Benzaazoua, M.; Hakkou, R.; Mansori, M. Coal Mine Wastes Recycling for Coal Recovery and Eco-Friendly Bricks Production. *Miner. Eng.* **2017**, *107*, 123–138. [CrossRef]
55. Nikvar-Hassani, A.; Hodges, R.; Zhang, L. Production of Green Bricks from Low-Reactive Copper Mine Tailings: Durability and Environmental Aspects. *Constr. Build. Mater.* **2022**, *337*, 127571. [CrossRef]
56. Lam, E.J.; Zetola, V.; Ramírez, Y.; Montofré, Í.L.; Pereira, F. Making Paving Stones from Copper Mine Tailings as Aggregates. *Int. J. Environ. Res. Public Health* **2020**, *17*, 2448. [CrossRef]
57. Xu, F.; Wang, S.; Li, T.; Liu, B.; Li, B.; Zhou, Y. Mechanical Properties and Pore Structure of Recycled Aggregate Concrete Made with Iron Ore Tailings and Polypropylene Fibers. *J. Build. Eng.* **2021**, *33*, 101572. [CrossRef]
58. Stevanović, D.; Karanac, M.; Ilić, M.; Miletić, S. Fly Ash from Coal Power Plants and its Application—A Step Towards Green Building. *Met. Mater. Data* **2024**, *1*, 117–123. [CrossRef]
59. Sathiparan, N.; Jaasim, J.H.M.; Banujan, B. Sustainable Production of Cement Masonry Blocks with the Combined Use of Fly Ash and Quarry Waste. *Materialia* **2022**, *26*, 101621. [CrossRef]
60. Siddique, R. Performance Characteristics of High-Volume Class F Fly Ash Concrete. *Cement Concr. Res.* **2004**, *34*, 487–493. [CrossRef]
61. Ghosh, S.K.; Kumar, V. (Eds.) *Circular Economy and Fly Ash Management: 2020 Overview*; Springer: Kolkata, India, 2020. [CrossRef]
62. European Commission; Eurostat. Waste Generation. [Data Set]. Available online: https://ec.europa.eu/eurostat/databrowser/view/env_wasgen/default/table?lang=en (accessed on 20 February 2025).
63. Ministry of Environmental Protection of the Republic of Serbia. *Waste Management Program in the Republic of Serbia for the Period 2022–2031*; Ministry of Environmental Protection: Belgrade, Serbia, 2022. (In Serbian)
64. Statistical Office of the Republic of Serbia. *Waste and Hazardous Chemicals*; Statistical Office of the Republic of Serbia: Belgrade, Serbia, 2023. Available online: https://www.stat.gov.rs/en-us/oblasti/zivotna-sredina/otpad-i-opasne-hemikalije/ (accessed on 20 January 2025).
65. Biznis i Finansije. *Kružna Ekonomija: Nije Svako Đubre za Bacanje*; Biznis i Finansije: 2020; Volume 178. BIF Press, Belgrade, Serbia. Available online: https://bif.rs/2020/10/biznis-finansije-178-kruzna-ekonomija-nije-svako-dubre-za-bacanje/ (accessed on 30 January 2025).
66. Jelena, D.; Marinkovic, S.; Ignjatovic, I.; Tosic, N. Concrete Based on Alkali Activated Fly Ash from One Power Plant in Serbia. *Int. J. Res. Eng. Technol.* **2014**, *3*, 1–9.
67. Alivojvodic, V.; Petronijevic, N. Position of copper within urban mining—Recovering potential from mine tailings. In Proceedings of the XV International Mineral Processing and Recycling Conference, Belgrade, Serbia, 17–19 May 2023; pp. 309–315.
68. Adewuyi, S.O.; Anani, A.; Luxbacher, K. Advancing Sustainable and Circular Mining Through Solid-Liquid Recovery of Mine Tailings. *Process Saf. Environ. Prot.* **2024**, *189*, 31–46. [CrossRef]
69. Urošević, S.; Vuković, M.; Pejčić, B.; Štrbac, N. Mining-Metallurgical Sources of Pollution in Eastern Serbia and Environmental Consciousness. *Int. J. Environ. Pollut.* **2018**, *34*, 103–115. [CrossRef]
70. Maksimović, M.; Bugarin, M.; Stevanović, Z.; Marinković, V. *Technogenic Deposit in the Area of the Old Flotation Tailing Dump in Bor (Field 1 and Field 2)*; Mining and Metallurgy Institute Bor: Bor, Serbia, 2014; Volume 16.
71. Pantović, R. Potencijal Tehnogenih Ležišta Bakra i Zlata (Flotaciona Jalovina) u Rudnom Kompleksu Bor-Srbija, 2021. Available online: https://mibor.rs/wp-content/uploads/2021/12/Analiza-rudnickog-otpda-u-Boru-RP.pdf (accessed on 19 January 2025).
72. Vukas, R.B. Mineralne Sirovine Srbije na Listi EU CRM-2020 Tokom Pandemije Covid-19 u Harmonizaciji sa Kodovima UNFC-2009 i Pojmovima PERC Standarda. *Tehnika* **2023**, *73*, 295–309. [CrossRef]
73. Soković, J.M.; Stanojlović, R.; Andrić, L.; Štirbanović, Z. Flotation Studies of Copper Ore Majdanpek to Enhance Copper Recovery and Concentrate Grade with Different Collectors. *J. Min. Metall. Sect. A Min.* **2019**, *55*, 53–65. [CrossRef]
74. Markovic, M. €2 Million EU-Funded Database of Serbian Mining Waste Incomplete. Analysis, Balkan Insight, 2024. Available online: https://balkaninsight.com/2024/10/22/e2-million-eu-funded-database-of-serbian-mining-waste-incomplete/ (accessed on 29 January 2025).
75. Lilić, N. *Studija o Proceni Uticaja na Životnu Sredinu Projekta Eksploatacije Rude Olova i Cinka "Vučkovog Ležišta" i Ležišta "Kula" u Sklopu Rudnika "Grot" A.D.—Kriva Feja*; Rudarsko-geološki fakultet, Univerzitet u Beogradu: Beograd, Serbia, 2019.
76. Antonijević, M.M.; Dimitrijević, M.D.; Stevanović, Z.O.; Šerbula, S.M.; Bogdanović, G.D. Investigation of the Possibility of Copper Recovery from the Flotation Tailings by Acid Leaching. *J. Hazard. Mater.* **2008**, *158*, 23–34. [CrossRef]
77. Stevanović, Z.; Antonijević, M.; Bogdanović, G.; Bugarin, M. The Effect of Oxidants Through a Tailing Dump Depth and the Leaching of Copper. *Carpathian J. Earth Environ. Sci.* **2013**, *8*, 29–38.
78. Stanković, S.; Morić, I.; Pavić, A.; Vojnović, S.; Vasiljević, B.; Cvetković, V. Bioleaching of Copper from Samples of Old Flotatio Tailings (Copper Mine Bor, Serbia). *J. Serb. Chem. Soc.* **2015**, *80*, 391–405. [CrossRef]
79. Sokić, M.D.; Milošević, V.D.; Stanković, V.D.; Matković, V.L.; Marković, B.R. Acid Leaching of Oxide-Sulfide Copper Ore Prior the Flotation: A Way for an Increased Metal Recovery. *Chem. Ind.* **2015**, *69*, 453–458. [CrossRef]

80. Stevanović, Z.; Antonijević, M.; Jonović, R.; Avramović, L.; Marković, R.; Bugarin, M.; Trujić, V. Leach-SX-EW Copper Revalorization from Overburden of Abandoned Copper Mine Cerovo, Eastern Serbia. *J. Min. Metall. Sect. B Metall.* **2009**, *45*, 45–57. [CrossRef]
81. Stanković, V.; Milošević, V.; Milićević, D.; Gorgievski, M.; Bogdanović, G. Reprocessing of the Old Flotation Tailings Deposited on the Bor Tailings Pond—A Case Study. *Chem. Ind. Chem. Eng. Q.* **2018**, *24*, 334–344. [CrossRef]
82. European Commission: Directorate-General for Internal Market, Industry; Entrepreneurship and SMEs; Grohol, M.; Veeh, C. *Study on the Critical Raw Materials for the EU 2023—Final Report*; Publications Office of the European Union: Luxembourg, 2023. [CrossRef]
83. European Commission, Joint Research Centre (JRC). *Best Available Techniques (BAT) Reference Document for Large Combustion Plants (LCP BREF)*; Publications Office of the European Union: Luxembourg, 2017. Available online: https://eippcb.jrc.ec.europa.eu/sites/default/files/2019-11/JRC_107769_LCPBref_2017.pdf (accessed on 16 February 2025).
84. Energija Balkana. Upotrebom Pepela, Šljake i Gipsa Štedi se Novac. Available online: https://energijabalkana.net/upotrebom-pepela-sljake-i-gipsa-stedi-se-novac/ (accessed on 20 January 2025).
85. Vukicevic, M.; Popović, Z.; Despotovic, J.; Lazarević, L. Fly Ash and Slag Utilization for the Serbian Railway Substructure. *Transport* **2016**, *33*, 1–10. [CrossRef]
86. Ministry of Environmental Protection; Environmental Protection Agency. *Report on the State of the Environment in the Republic of Serbia for 2023*; Ministry of Environmental Protection: Belgrade, Serbia, 2024; (In Serbian). Available online: https://sepa.gov.rs/wp-content/uploads/2024/12/ZivotnaSredina2023.pdf (accessed on 20 February 2025).
87. Radovanović, D.Đ.; Kamberović, Ž.J.; Korać, M.S.; Rogan, J.R. Solidified structure and leaching properties of metallurgical wastewater treatment sludge after solidification/stabilization process. *J. Environ. Sci. Health Part A* **2016**, *51*, 34–43. [CrossRef]
88. Štulović, M.; Radovanović, D.; Dikić, J.; Gajić, N.; Djokić, J.; Kamberović, Ž.; Jevtić, S. Utilization of Copper Flotation Tailings in Geopolymer Materials Based on Zeolite and Fly Ash. *Materials* **2024**, *17*, 6115. [CrossRef]
89. Đolić, M.; Karanac, M.; Radovanović, D.; Umicević, A.; Kapidzić, A.; Velicković, Z.; Marinković, A.; Kamberović, Z. Closing the loop: As(V) adsorption onto goethite impregnated coalcombustion fly ash as integral building materials. *J. Clean. Prod.* **2021**, *303*, 126924. [CrossRef]
90. Petronijević, N.; Radovanović, D.; Štulović, M.; Sokić, M.; Jovanović, G.; Kamberović, Ž.; Stanković, S.; Stopić, S.; Onjia, A. Analysis of the Mechanism of Acid Mine Drainage Neutralization Using Fly Ash as an Alternative Material: A Case Study of the Extremely Acidic Lake Robule in Eastern Serbia. *Water* **2022**, *14*, 3244. [CrossRef]
91. Petronijević, N.; Stanković, S.; Radovanović, D.; Sokić, M.; Marković, B.; Stopić, S.R.; Kamberović, Ž. Application of the Flotation Tailings as an Alternative Material for an Acid Mine Drainage Remediation: A Case Study of the Extremely Acidic Lake Robule (Serbia). *Metals* **2020**, *10*, 16. [CrossRef]
92. Law on Environmental Protection. Official Gazette of the Republic of Serbia, No. 135/2004-29, 36/2009-144, 36/2009-115, 72/2009-164, 43/2011-88, 14/2016-3, 76/2018-3, 95/2018-267, 94/2024-391. (In Serbian). Available online: https://www.paragraf.rs/propisi/zakon_o_zastiti_zivotne_sredine.html (accessed on 10 January 2025).
93. Law on Waste Management. Official Gazette of the Republic of Serbia, No. 36/2009, 88/2010, 14/2016, 95/2018, 35/2023. (In Serbian). Available online: https://www.paragraf.rs/propisi/zakon_o_upravljanju_otpadom.html (accessed on 10 January 2025).
94. Law on Water. Official Gazette of the Republic of Serbia, No. 30/2010, 93/2012, 101/2016, 95/2018. (In Serbian). Available online: https://www.paragraf.rs/propisi/zakon_o_vodama.html (accessed on 10 January 2025).
95. Air Protection Law. Official Gazette of the RS, no. 36/2009, 10/2013, 26/2021. (In Serbian). Available online: https://www.paragraf.rs/propisi/zakon_o_zastiti_vazduha.html (accessed on 10 January 2025).
96. Law on Mining and Geological Research. Official Gazette of the RS, no. 101/2015, 95/2018, 40/2021. (In Serbian). Available online: https://gras.org.rs/wp-content/uploads/2017/10/law-on-mining-and-geological-explorations-of-serbia-2015.pdf (accessed on 10 January 2025).
97. Rule Book on Categories, Testing and Classification of Waste. Official Gazette of the RS, no. 56/2010, 93/2019, 39/2021, 65/2024. (In Serbian). Available online: https://www.paragraf.rs/propisi/pravilnik-o-kategorijama-ispitivanju-i-klasifikaciji-otpada.html (accessed on 10 January 2025).
98. Rule Book of Environmental Protection of the Republic of Serbia. Rulebook on Types of Waste for Which a Request Can Be Submitted, Permitted Procedures and Treatment Technologies for Types of Waste and Other Special Elements for Determining the Termination of Waste Status; "Official Gazette of RS", No. 19/2024, 47/2024. (In Serbian). Available online: https://www.ekologija.gov.rs/propisi/pravilnici (accessed on 10 January 2025).
99. Rulebook on Technical Requirements and Other Special Criteria for Certain Types of Waste That Cease to Be Waste. Official Gazette of the RS, no. 78/2019. (In Serbian). Available online: https://www.pravno-informacioni-sistem.rs/SlGlasnikPortal/eli/rep/sgrs/ministarstva/pravilnik/2019/78/1 (accessed on 10 January 2025).
100. Directive 1999/31—Landfill of Waste. Available online: https://eur-lex.europa.eu/eli/dir/1999/31/oj/eng (accessed on 20 December 2024).

101. Directive 2006/21/EC. Directive on the Management of Waste from Extractive Industries. Official Journal of the European Union 2006. Available online: https://eur-lex.europa.eu/eli/dir/2006/21/oj/eng (accessed on 16 December 2024).
102. Directive 2008/98/EC. Waste Framework Directive. Official Journal of the European Union 2008. Available online: https://eur-lex.europa.eu/eli/dir/2008/98/oj/eng (accessed on 16 December 2024).
103. European Commission. Communication from the Commission to the Council and the European Parliament—The Raw Materials Initiative: Meeting Our Critical Needs for Growth and Jobs in Europe. EUR-Lex 2008, Document 52008DC0699. Available online: https://eur-lex.europa.eu/legal-content/EN/TXT/?uri=CELEX:52008DC0699 (accessed on 16 December 2024).
104. Directive 2010/75/EU. Directive on Industrial Emissions (Integrated Pollution Prevention and Control). Official Journal of the European Union 2010. Available online: https://eur-lex.europa.eu/eli/dir/2010/75/oj/eng (accessed on 16 December 2024).
105. Directive 2000/60/EC. Water Framework Directive. Official Journal of the European Union 2000. Available online: https://eur-lex.europa.eu/eli/dir/2000/60/oj/eng (accessed on 16 December 2024).
106. Directive 2014/95/EU. Non-Financial Reporting Directive (NFRD). Official Journal of the European Union 2014. Available online: https://eur-lex.europa.eu (accessed on 16 December 2024).
107. *EN 12620:2002+A1:2008*; Aggregates for Concrete. European Committee for Standardization: Brussels, Belgium, 2008.
108. European Commission. *The European Green Deal*; COM(2019) 640 final; European Commission: Brussels, Belgium, 2019.
109. European Commission. *Communication from the Commission to the European Parliament, the Council, the European Economic and Social Committee and the Committee of the Regions: A New Circular Economy Action Plan for a Cleaner and More Competitive Europe*; COM(2020) 98 final; European Commission: Brussels, Belgium, 2020.
110. Regulation (EU) 2024/3110 of the European Parliament and of the Council of 27 November 2024 Laying Down Harmonised Rules for the Marketing of Construction Products and Repealing Regulation (EU) No 305/2011 (Text with EEA Relevance). Off. J. Eur. Union 2024, L 3110, 18.12.2024. Available online: https://eur-lex.europa.eu/eli/reg/2024/3110/oj/eng (accessed on 16 January 2025).
111. *Best Available Techniques (BAT) Reference Document for the Management of Waste from Extractive Industries, in Accordance with Directive 2006/21/EC*; EUR 28963 EN; Publications Office of the European Union: Luxembourg, 2018; ISBN 978-92-79-77178-1. Available online: https://ec.europa.eu/environment/pdf/waste/mining/MWEI%20BREF.pdf (accessed on 16 January 2025).
112. *ASTM C618-22*; Standard Specification for Coal Fly Ash and Raw or Calcined Natural Pozzolan for Use in Concrete. ASTM International: West Conshohocken, PA, USA, 2022.
113. Statista. Cement Decarbonization Costs Worldwide 2023 by Approach. Available online: https://www.statista.com/statistics/1557813/cement-decarbonization-costs-worldwide-by-approach/ (accessed on 21 February 2025).
114. EPS. Sales of Dry Fly Ash and Landfill Ash (Waste Slag) for 2023. (In Serbian). Available online: https://www.eps.rs/cir/tent/oglasi/Pages/Prodaja-suvog-lete%C4%8Deg-pepela-i-pepela-sa-deponije-(otpadne-%C5%A1ljake)-za-2023.godinu.aspx (accessed on 20 February 2025).
115. ES2536464B1. Process for the Inertization of Ash from the Combustion of Municipal Solid Waste. Available online: https://patents.google.com/patent/ES2536464B1/en (accessed on 16 February 2025).
116. Velásquez, J.R.; Schwartz, M.; Phipps, L.M.; Restrepo-Baena, O.J.; Lucena, J.; Smits, K.M. A review of the environmental and health implications of recycling mine tailings for construction purposes in artisanal and small-scale mining communities. *Extr. Ind. Soc.* **2022**, *9*, 101019. [CrossRef]
117. ISO. The Circular Economy: Building Trust through Conformity Assessment. Available online: https://www.iso.org/sectors/environment/circular-economy (accessed on 20 February 2025).

minerals

MDPI

Article

Utilisation of Zinc Processing Tailings (ZPTs) in Fired Clay Bricks Manufacturing: Case Study

Ali Messai [1,2], Juan M. Menéndez-Aguado [3], Linda Berrekbia [2], Ikram Meramria [2], Begoña Fernández Pérez [3], Vladimir Nikolić [4,*], Milan Trumić [4] and Amir Boustila [1,2]

1 Laboratory of Applied Research in Engineering Geology, Geotechnics, Water Sciences & Environment, Ferhat Abbas University Setif 1, Setif 19137, Algeria; a.messai@ensti-annaba.dz (A.M.); a.boustila@ensti-annaba.dz (A.B.)
2 National Higher School of Technology and Engineering, Department of Mining, Metallurgy and Materials Engineering, Annaba 23005, Algeria; lindaberrekbiaensti@gmail.com (L.B.); ikrammeramria176@gmail.com (I.M.)
3 Asturias Raw Materials Institute, University of Oviedo, Gonzalo Gutiérrez Quirós, 33600 Mieres, Spain; maguado@uniovi.es (J.M.M.-A.); fernandezbegona@uniovi.es (B.F.P.)
4 Technical Faculty in Bor, University of Belgrade, 19210 Bor, Serbia; mtrumic@tfbor.bg.ac.rs
* Correspondence: vnikolic@tfbor.bg.ac.rs

Abstract: Zinc processing tailings (ZPTs) of the Kharzet Youcef processing complex, Setif, Algeria, are mainly stockpiled in tailing dumps without use, occupying significant surfaces and negatively influencing the human environment and health. Incorporating ZPTs into building materials manufacturing is an effective solution to meet the dual objectives of environmental protection and economic development. This study investigates the influence of firing temperature and integrating ZPTs as a partial replacement for clay on the physic-mechanical properties of fired clay bricks (FCBs). Microstructural, chemical, and mineralogical analyses of ZPTs and clay were carried out by SEM-EDS, XRF, and XRD, respectively. Seven mixtures were produced with various percentages of ZPTs added to clay (0%, 5%, 10%, 15%, 20%, 25%, and 30%) and were fired at two different temperatures (900 and 1000 °C) at a ramp rate of 5 °C. Physic-mechanical tests were carried out on different brick specimens, and the results obtained showed that the FCBs incorporated with 10% of ZPTs produced the highest flexural strength of 6.24 MPa, compressive resistance of 29.78 MPa, bulk density of 1.37 g/cm^3, and water absorption of 15.1% at 900 °C. Therefore, the recycling of ZPTs for FCBs manufacturing is feasible and an effective alternative waste disposal solution for sustainable development while reducing negative environmental impacts.

Keywords: zinc processing tailings (ZPTs); fired clay bricks (FCBs); construction; Kharzet Youcef; physic-mechanical tests

1. Introduction

Non-renewable mineral resources are the basic building blocks supporting global economic and social growth [1]. The exponential growth in mining activities has generated large amounts and various types of wastes. These wastes are mainly in the form of tailings that are disposed in dumps without proper management, inevitably affecting the environment [2–4]. In this context, zinc mining and processing tailings are a primary source of toxic heavy metal discharge into the environment [1,2,5].

In this regard, the recycling and utilisation of mining wastes and tailings as alternative raw materials is regarded as an environmentally friendly approach to reduce their negative impact on the environment on the one hand and to prevent the depletion of non-renewable resources and improve the population's health through the proper and effective disposal of wastes on the other hand [6–8].

Furthermore, zinc wastes have a chemical composition similar to that of the natural raw materials used in the building and construction industries [9], so they can be used as a partial replacement for clay in bricks [10–12], cement [13–16], concrete admixture [17–19], mortar [20–22], the production of ceramics [23–26], and cementitious paste [27,28].

In other research [11], lead–zinc tailings were used as the primary raw material and combined with clay and fly ash as supplementary components to produce fired bricks. The optimal manufacturing conditions included a ratio of lead–zinc tailings to clay to fly ash equal to 6:3:1, a forming pressure of 20 MPa, a firing temperature of 1080 °C, and a holding time of 60 min. Under these conditions, the bricks attained a compressive strength of 34.94 MPa and a water absorption rate of 16.02%, meeting the requirements of the Chinese sintered ordinary bricks standard.

Lead–zinc tailings have also been utilised in the production of unfired bricks [29]. Lightweight unfired bricks were developed using lead–zinc tailings as fine aggregates, achieving a compressive strength of 9.3 MPa, making them suitable as building filler blocks. Similarly [30], unfired bricks incorporating lead–zinc tailings as a pozzolanic material were manufactured, with a compressive strength meeting the MU20 standard, reaching an average of 20 MPa. These studies demonstrate that unfired bricks made from lead–zinc tailings can attain acceptable mechanical properties.

In Algeria, Chaabet El Hamra deposit is the only zinc deposit in the country; it is operated by the National Company of Non-Ferrous Mining Products (ENOF). Its raw ore is extracted using the room and pillar method and directly conveyed to the mineral processing plant in order to increase its quality. Froth flotation is the method used to produce a final high-grade concentrate, while rejects are deposited in a dump. This dump occupies an area of 7 ha, having received residue issues from the treatment of Chaabet El Hamra ore since 1994.

These tailings have primarily been studied from an environmental perspective, focusing on heavy metal accumulation in species [31–33]. However, rejects have not been specifically examined for incorporation into fired clay bricks. This study follows a circular economy approach by evaluating the suitability of ZPTs as a partial replacement for clay in fired brick manufacturing in Algeria, aiming to explore their potential as a sustainable raw material. The novelty of this research will contribute to waste reduction and environmental sustainability. Through physical–mechanical testing and microstructural analysis of the produced brick specimens, the research aims to transform these industrial residues into a valuable resource. In addition to reducing environmental impact, this approach offers a sustainable alternative for the construction industry, with findings that will help to assess the potential of these materials and optimise their performance for large-scale applications.

2. Study Area Description

Geographic Location

The mineral processing plant of Kherzet Youcef ore (Chaabet El Hamra Mine) is located in eastern Algeria, 5 km northwest of Ain Azel, approximately 50 km southeast of the municipality of Setif (Figure 1).

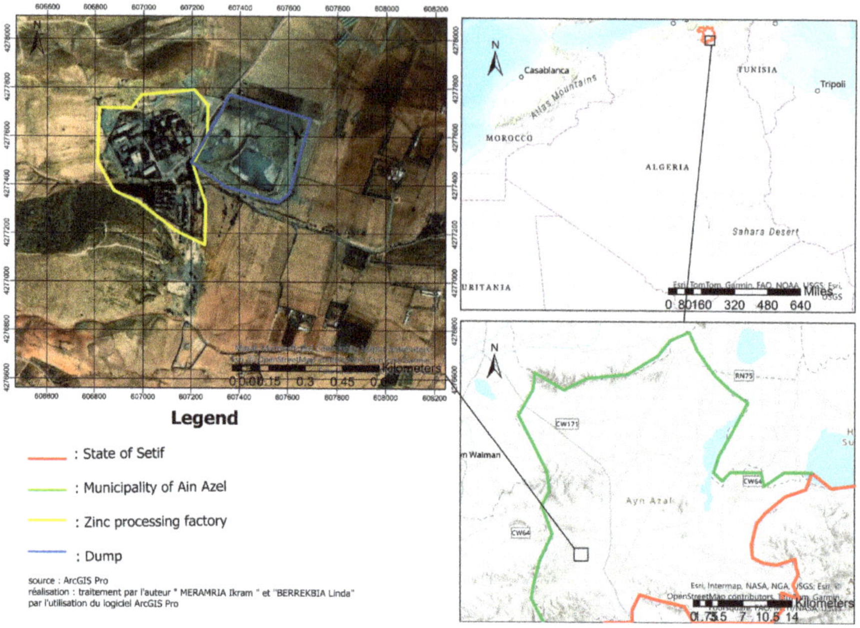

Figure 1. Geographic location of the study area. Red colour present the borders of Setif state.

3. Materials and Methods

3.1. Materials

The clay used in this study was obtained from the El Amel brick factory, El Taref, Algeria. The sampling procedure was applied in the discharge of a rolling mill (<1 mm). The sampling was performed using a steel shovel, avoiding any contamination of the sample on the one hand and ensuring that the sample taken reflected the composition and characteristics of the material used in the brick manufacturing process on the other hand.

The ZPT samples were obtained from five distinctive dump locations of the Kharzet Youcef zinc processing plant. Then, subsamples from each zone were homogenised and five samples were finally obtained. After that, the materials (clay and ZPTs) were dried at 105 °C for 24 h. ZPTs were ground using a planetary ball mill type Welte WTR 950 W/R-V and then sieved to select particles with an average diameter of less than 0.074 mm.

Five samples of ZPTs collected from the Kharzet Youcef dumps were thoroughly mixed and prepared by conning and quartering to prepare a single representative sample.

3.2. Moulding and Sintering of Brick Specimens

To manufacture bricks, ZPTs were properly mixed with clay in varying portions, as presented in Table 1. The different mixtures were first blended for 10 min in a blender, with 10% by mass of water in order to obtain semi-dry homogenous mixes. Homogenised mixtures were manually pressed using a mould with dimensions of 40 × 40 × 160 mm to produce nine (9) brick specimens for each combination. The obtained specimens were left in ambient air for 24 h, and then dried in an oven at 105 °C for 24 h to decrease their moisture content. The dried specimens were fired in a laboratory electrical furnace at a rate of 5 °C/min from room temperature to the desired firing temperature, where two different temperatures were targeted (900 °C and 1000 °C), and held for two hours at the desired firing temperature. After firing, the bricks were cooled to room temperature by natural convection inside the laboratory electrical furnace.

Table 1. Mass ratios of different raw materials.

Formulations	F0	F5	F10	F15	F20	F25	F30
ZPTs (%)	0	5	10	15	20	25	30
ZPTs (g)	0	15	30	45	60	75	70
Clay (%)	100	95	90	85	80	75	70
Clay (g)	300	285	270	255	240	225	210

The procedure followed for brick manufacturing at the laboratory is shown in Figure 2.

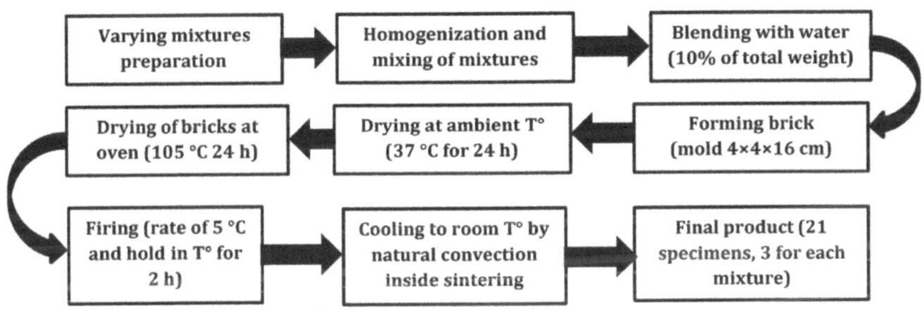

Figure 2. Schematic representation of brick specimens manufacturing, own elaboration.

3.3. Characterisation

The main chemical composition of the raw materials was determined by X-Ray Fluorescence spectroscopy (XRF PANalytical AXIOS minerals, Malvern Panalytical, Almelo, Netherlands) and Inductively Coupled Plasma Optical Emission spectroscopy (Agilent technologies 5900, Agilent Technologies, Santa Clara, CA, USA). The loss on ignition (LOI) was measured by heating the samples at 1000 °C for 2 h in a muffle furnace, followed by weighing the residue to determine the mass loss.

The mineralogical composition was investigated with X-Ray Diffractions using a device type X'pert Pro MRD (Malvern Panalytical, Almelo,The Netherlands).

The surface morphology and pore structure of the raw materials and sintered brick specimens were investigated by SEM model FEI Quanta 250 (FEI, Eindhoven, The Netherlands) equipped with an energy-dispersive X-ray spectroscopy (EDS) detector.

The plasticity and liquidity limits of the mixtures were determined due to ASTM D4318-17e1 [34], while linear shrinkage was measured according to ASTM D4943-18 [35]. The bulk density and water absorption values of the sintered samples were calculated using the Archimedes method according to ASTM C29/C29M-17a [36] and ASTM ASTM C373-18 [37], respectively. These parameters were determined using the following equations:

$$Bulk\ density\ (\rho) = \frac{M_{dry}}{V}$$

M_{dry} is the dry mass of the sample (g),
V is the volume of the sample (cm^3).

$$Water\ absorption\ (A) = \frac{M_{immerse} - M_{dry}}{M_{dry}} 100$$

$M_{immersed}$ is the mass of the sample after immersion (g),
M_{dry} is the dry mass of the sample (g).

Linear shrinkage (%) was determined from the lengths of the brick specimens after drying and after sintering treatment by a calliper with a precision of ±0.01 mm according to ASTM C326-09 [38].

$$Linear\ shrinkage\ (S) = \frac{L_{initial} - L_{final}}{L_{initial}} 100$$

$L_{initial}$ is the initial length of the sample (mm),

L_{final} is the length after thermal treatment (mm).

Compressive resistance was assessed using a mechanical testing machine 'Bera Test' (MaTest, Triviolo, Italy) with a capacity of 2000 kN; it can apply a maximum pressure equal to 618.17 bars. The surface area of the piston is 323.65 cm^2, according to ASTM C67-22 [39]. Flexural strength was measured using the mechanical testing machine Bera Test' with a capacity of 150 kN, which can apply a maximum pressure equal to 122.28 bars. The surface area of the piston is 122.71 cm^2, according to ASTM C674-88 [40].

$$The\ compressive\ resistance\ (\sigma_c) = \frac{F}{A}$$

where:

F is the maximum applied force (N),

A is the loaded surface area (mm^2).

$$The\ flexural\ strength\ (\sigma_c) = \frac{3FL}{2bh^2}$$

where:

F is the maximum applied load (N),

L is the span between supports (mm),

b is the width of the sample (mm),

h is the thickness of the sample (mm).

4. Results and Discussions

4.1. ZPTs and Clay Characterisation

4.1.1. Phases Identification

The results of the XRD analysis of clay (Figure 3) confirm the presence of kaolinite ($Al_2Si_2O_5(OH)_4$), quartz (SiO_2), dolomite ($CaMg(CO_3)_2$), hematite (Fe_2O_3), and fluorite (CaF_2). The dominant peak of quartz is consistent with the high SiO_2 content (42.9%), indicating its abundance in the sample. The presence of kaolinite aligns with the significant Al_2O_3 content (23.3%), which plays a crucial role in ceramic phase formation during firing. Hematite is supported by the total iron (Fe_T) content (8.25%), contributing to fired bricks' colouration changes. Dolomite, identified by its peaks, corresponds to the CaO (8.9%) and MgO (4.1%) content, and its decomposition during heating can influence porosity and densification.

The X-ray diffraction (XRD) results of the ZPTs presented in Figure 4 indicate the presence of dolomite ($CaMg(CO_3)_2$), quartz (SiO_2), gypsum ($CaSO_4{\cdot}2H_2O$), and pyrite (FeS_2).

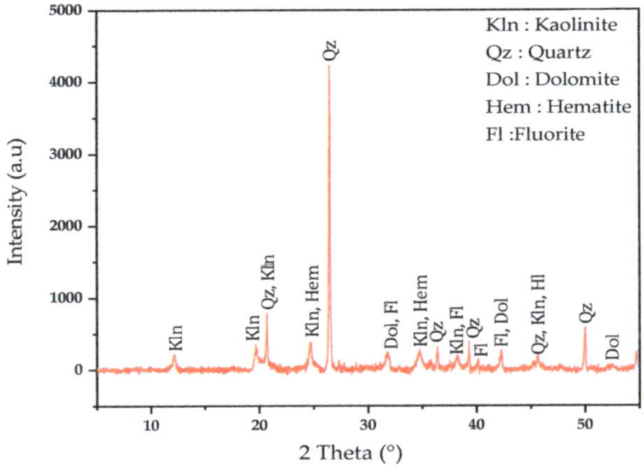

Figure 3. X-ray diffraction patterns of clay.

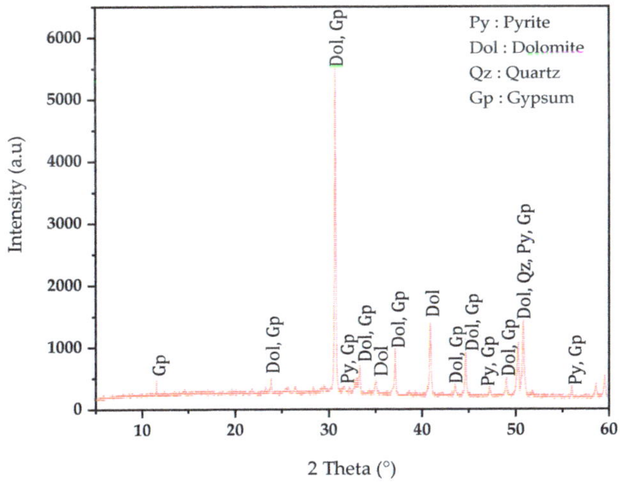

Figure 4. XRD phase's identification of Kharzet Youcef ZPTs.

4.1.2. Chemical Composition

Chemical analysis of the clay used in this study (Table 2) reveals a composition with a high content of SiO_2 (42.9%) and Al_2O_3 (23.3%), accompanied by significant levels of total iron (Fe_T = 8.25%) and CaO (8.9%).

Table 2. Chemical composition of clay.

Elements	SiO_2	Al_2O_3	Fe_T	CaO	K_2O	MgO	TiO_2	Na_2O	LOI
Percentage (%)	42.9	23.3	8.25	8.9	2.8	4.1	2.7	1.1	5.95

The percentage of zinc in the five collected samples was quantified and the results are presented in Table 3. A variation in zinc concentrations is shown between the different areas of the site. Zone 4 has the highest concentration of zinc (1.01%), while zone 3 has the lowest concentration (0.23%). This variation may be due to differences in the treatment process or in local geology.

Table 3. Zinc content in collected samples of ZPTs.

Zone of Sample	1	2	3	4	5
Zn (%)	0.91	0.71	0.23	1.01	0.65

The main components of the ZPTs of the composite sample are presented in Table 4. The sample is primarily composed of CaO (31.62%) and exhibits a high loss on ignition (LOI) of 35.46%, which suggests the presence of significant volatile components and/or decomposable phases. This elevated LOI value can be attributed to the thermal decomposition of carbonates (such as calcium carbonate, $CaCO_3$, and magnesium carbonate, $MgCO_3$), leading to the release of CO_2. SiO_2 and MgO are also found in significant quantities, representing 9.45% and 10.74%, respectively.

Table 4. Chemical composition of Kharzet Youcef ZPTs.

Elements	SiO_2	Al_2O_3	Fe_2O_3	CaO	K_2O	MgO	MnO	Na_2O	P_2O_5	S	LOI
Percentage (%)	9.45	2.2	5.05	31.62	0.38	10.74	0.08	0.54	0.05	3.18	35.46

The percentages of heavy metals are shown in Table 5. Zinc, lead, and strontium are present in notable quantities, indicating potential environmental and health risks linked to the dispersion of these rejects.

Table 5. Heavy metals content of Kharzet Youcef ZPTs.

Elements	Ti	Zn	Pb	Cu	Sr
Percentage (%)	0.00918	0.04460	0.04171	0.00163	0.00293

In conclusion, the silica and alumina content in ZPTs can qualify them to partially replace clay in bricks manufacturing.

4.1.3. Microstructure

- Clay:

The SEM image (Figure 5) shows a heterogeneous microstructure mixed with compact and porous regions. The appearance of visible particles of varying sizes (from a few micrometres to around 100 µm) and irregular shapes indicates a complex mixture of different mineral phases. Angular particles corresponding to quartz are identified in the XRD results.

There are several visible pores and voids within the clay matrix. Their distribution and size suggest that the clay has a certain porosity level, which can affect its behaviour during the drying and firing processes in brick manufacturing.

EDS analysis (Table 6) confirms the presence of oxygen (O), magnesium (Mg), aluminium (Al), silicon (Si), chlorine (Cl), potassium (K), calcium (Ca), titanium (Ti), and iron (Fe). These results are consistent with the XRD findings, which identified quartz (SiO_2), kaolinite ($Al_2Si_2O_5(OH)_4$), siderite ($FeCO_3$), and halite (NaCl). Oxygen, silicon, and aluminium are fundamental to the clay structure, forming part of quartz and kaolinite. Magnesium and calcium likely originate from minor amounts of carbonate minerals. Iron, present in siderite, is key in influencing the brick's colour, imparting a reddish hue while enhancing its strength and hardness. Notably, potassium (K) and sodium (Na), also identified in the sample, act as fluxing agents during the sintering process, lowering the firing temperature, promoting densification, and contributing to improved mechanical properties. The minor presence of titanium and chlorine suggests impurities that must be managed.

Figure 5. SEM-EDS results of clay.

Table 6. EDS results of clay-scanned zone.

Elements	O	Mg	Al	Si	Cl	K	Ca	Ti	Fe
% of mass	58.68	1.70	12.57	20.86	0.51	1.32	0.55	0.47	3.34

- ZPTs:

The image obtained by SEM of the ZPTs (Figure 6) shows a heterogeneous granular structure composed of various mineral phases. The EDS analysis (Table 7) reveals a composition mainly dominated by oxides, carbonates, sulphates, and silicates, reflecting the mineralogical diversity of zinc processing discharges. The presence of sulphur and iron could indicate phases such as pyrite, while calcium and magnesium are consistent with dolomite and gypsum.

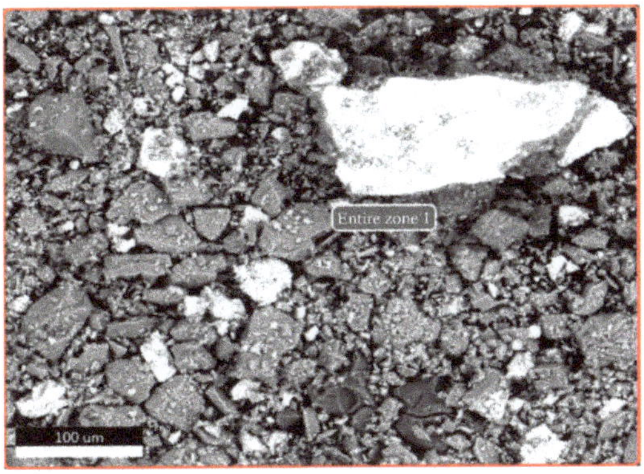

Figure 6. SEM-EDS results of ZPTs.

Table 7. EDS results of ZPTs.

Elements	O	Mg	C	Al	Si	S	Ca	Zn	Fe
% of mass	58.07	9.33	17.12	2.97	1.84	3.77	4.57	0.13	2.2

- Bricks:

SEM-EDS was carried out on bricks made with clay and ZPTs with the best mechanical resistance values. The analysis of the microstructure of the briquette prepared with 100%

clay and fired at 950 °C, as presented in Figure 7, shows a porous and heterogeneous texture, characterised by dense zones and cavities. The EDS analysis (Table 8) indicates a majority composition of oxygen (56.13%), silicon (18.50%), aluminium (10.84%), and iron (11.26%). The significant presence of iron may indicate a contribution to the typical reddish colouration of fired bricks. These results agree with the XRD analysis, which identified dominant crystalline phases such as quartz, kaolinite, and siderite, while XRF confirmed their elemental composition.

Figure 7. SEM results of brick prepared with 100% clay and fired at 900 °C.

Table 8. EDS results of brick prepared with 100% clay and fired at 900 °C.

Elements	O	Mg	K	Al	Si	Ca	Ti	Fe
% of mass	56.13	1.88	0.82	10.84	18.5	0.32	0.24	11.26

According to [41], kaolinite dehydroxylates around 500–600 °C, leading to the formation of an amorphous metakaolinite phase, which subsequently transforms into mullite (~950–1000 °C), a key contributor to the mechanical strength of bricks. Similarly, siderite decomposes at approximately 600 °C, releasing CO_2 and forming hematite (Fe_2O_3), which enhances the reddish colouration of bricks and contributes to densification. Improvements in mechanical strength and are is primarily attributed to the reduction in porosity due to sintering and the presence of quartz and mullite, which provide a rigid structural framework embedded in a glassy matrix. These findings are consistent with previous research on the mineralogical evolution of fired clay bricks.

These phases are typical of bricks fired at high temperatures, giving the briquettes their mechanical and durability properties.

The SEM image of the clay briquette with 30% ZPTs (Figure 8) shows a heterogeneous structure, with grains and pores of varying sizes, which can influence the mechanical properties and durability of briquettes. The EDS results (Table 9) confirm the presence of dominant elements (O, Mg, Al, Si, Ca, and Fe) characteristic of clay and ZPTs. The appearance of zinc (1.51%) and the augmentation of Mg, K, Ti, and Ca in the analysis of this briquette confirm the good incorporation of ZPTs into the matrix.

Figure 8. SEM results of brick prepared with 10% of ZPTs and fired at 900 °C.

Table 9. EDS results of bricks made with 30% ZPTs and fired at 900 °C.

Elements	O	Mg	Al	Si	Mo	S	Pb	K	Ca	Ti	Fe	Zn
% of mass	52.68	4.81	7.73	13.09	0.85	2.3	0.65	0.76	9.45	0.39	5.05	1.51

According to [2,12], the heavy metals in waste residues and sludge become stable in raw materials after irreversible transformation into mineral phases by thermal processes (brick sintering), and metal leaching is inhibited dramatically. These thermal treatments not only offer a cost-effective method for producing construction materials incorporating tailings, but also contribute to the stabilisation of heavy metals within the newly fabricated materials. Similar to the adsorption mechanisms observed in zeolite-based materials, the thermal treatment of ZPT-containing bricks may facilitate the immobilization of heavy metals by incorporating them into stable mineral phases, thus reducing their leaching potential [42].

4.1.4. Mapping

Mapping was carried out on bricks made with clay and ZPTs with the best mechanical resistance values. The results are shown in Figures 9 and 10.

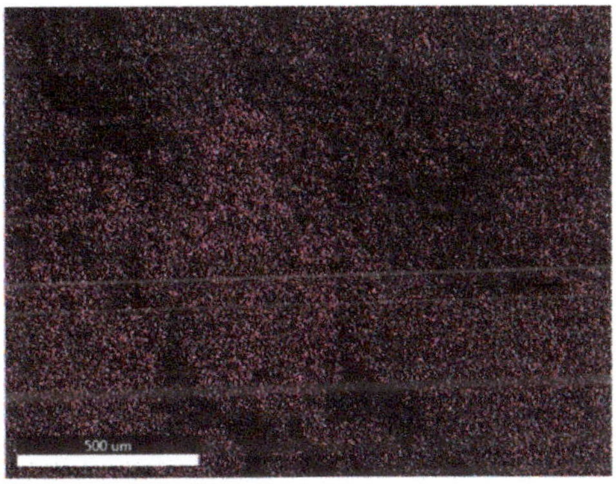

Figure 9. SEM cartography of briquette made with 100% clay and fired at 900 °C.

Figure 10. SEM cartography of briquette made with 10% ZPTs and fired at 900 °C.

- Briquette of 100% clay:

SEM mapping (Figure 9) shows a homogeneous structure of the briquette with specific inclusions detectable by the localised distribution of elements like silicon, oxygen, and aluminium, corresponding to their high percentages in the chemical composition of the clay. Silicon (purple) and oxygen (red) cover the surface evenly, while aluminium (sky blue) is also well distributed. Iron (in blue), on the other hand, is present in a more localised manner, indicating specific iron-rich inclusions or phases. Magnesium (in orange) and potassium (in mustard green) appear in lower quantities, reflecting their lower proportion in the clay.

- Briquette of 10% Zn-900 °C:

The SEM mapping image shows the distribution of elements in a briquette manufactured by integrating 10% of zinc waste with clay and firing at 900 °C. The result shows that the integration of ZPTs did not disturb the uniform distribution of essential elements, including aluminium, silica, iron, and oxygen, showing a good compatibility between the two components (ZPTs and clay) and the good integration of ZPTs into the clay matrix. The presence of sulphur indicates its introduction via ZPTs.

Magnesium and calcium, present in high concentrations in zinc processing tailings (CaO = 31.62% and MgO = 10.74%), play a crucial role in the sintering process. These elements act as fluxing agents, promoting vitrification at lower firing temperatures, enhancing the bricks' density and mechanical strength while reducing overall porosity. This can lead to improved water and frost resistance, reduced drying shrinkage, and minimised warping during firing. Additionally, magnesium contributes to high-temperature strength, making it particularly beneficial for refractory applications. However, despite these advantages, the decomposition of magnesium- and calcium-bearing phases can also generate significant gas release during firing, increasing brick porosity. This gas-induced porosity can compromise mechanical properties and water resistance, counteracting the densification effects of sintering. The observed increase in water absorption and reduction in mechanical strength after ZPT addition suggest that gas release plays a dominant role in determining the final properties of the bricks. Therefore, careful control of the firing temperature and atmosphere is essential to balance these competing effects and optimise brick performance.

4.1.5. Profilometry 3D

The surface roughness of bricks made with clay and ZPTs with the best mechanical resistance values was measured. The results are shown in Figures 11–14.

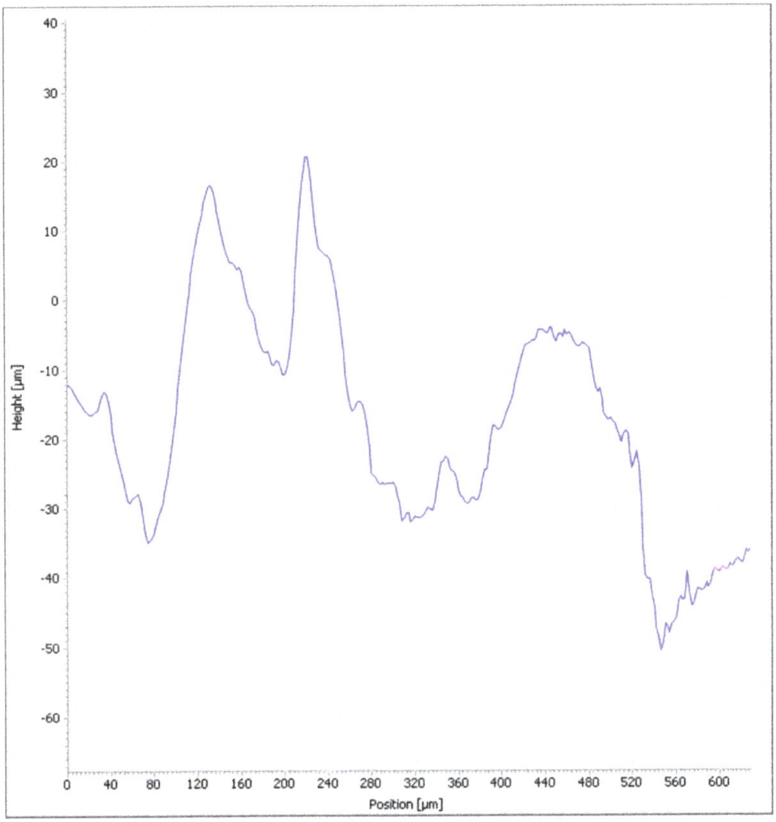

Figure 11. Surface roughness measurement results of briquette made with 100% clay: Cross-section profile curve.

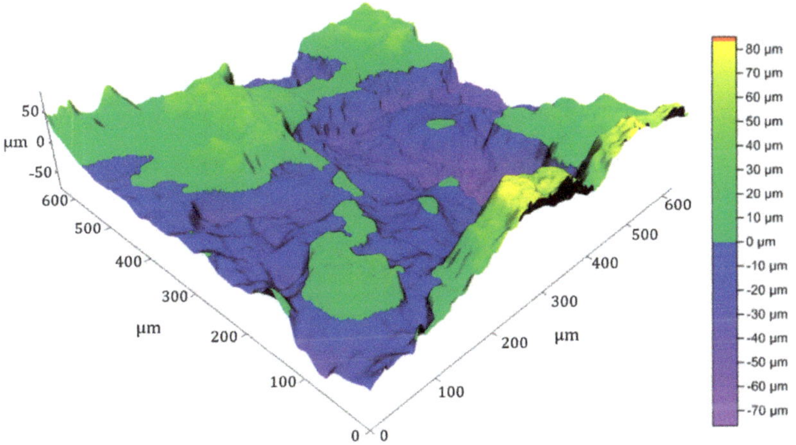

Figure 12. Surface roughness measurement results of briquette made with 100% clay: Surface topography.

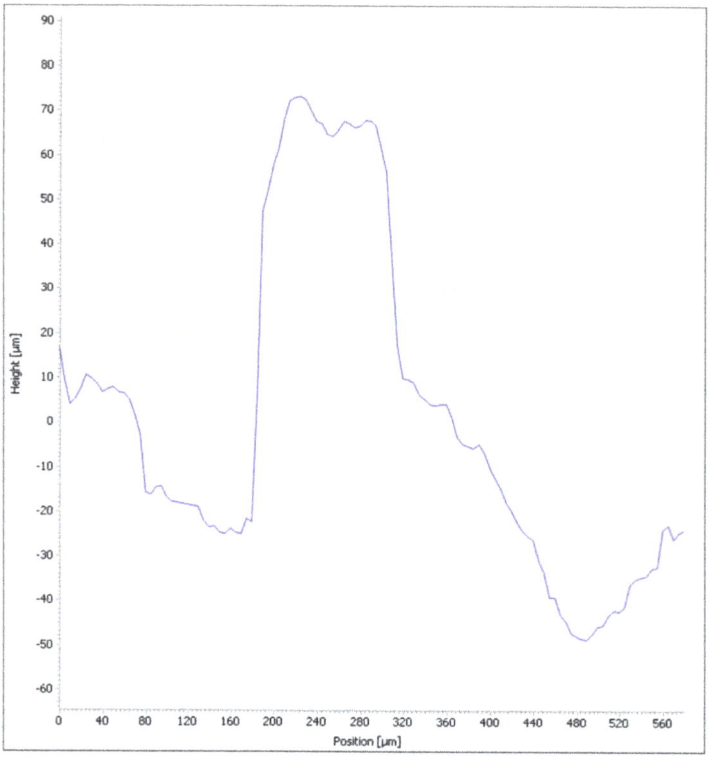

Figure 13. Surface roughness results of bricks made with 10% ZPTs and fired at 900 °C: Cross-section profile curve.

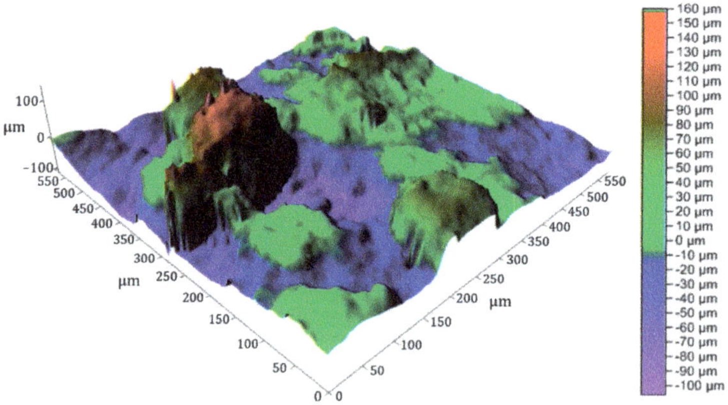

Figure 14. Surface roughness results of bricks made with 10% ZPTs and fired at 900 °C: Surface topography.

- Briquette of 100% clay:

The curve representing the height variation as a function of position for a profile line of the briquette (Figure 11) shows significant fluctuations in this line, with variations ranging from −60 µm to approximately 30 µm. This line's average roughness (Ra) is 4.42 µm, while Rq = 5.525 µm and Rz = 17.68 µm, corresponding to a moderate roughness.

The 3D image of the clay briquette surface (Figure 12) reveals more detailed height variations, confirming the irregularities observed in the profile line curve on the same surface and that the briquette is specified by a moderate roughness, with peaks and relatively pronounced troughs.

- Briquette of 10% ZPTs fired at 900 °C:

The results of measuring the roughness of the briquette with the best physic-mechanical properties (10% of ZPTs added to the clay and fired at 900 °C) are presented in Figures 13 and 14.

The linear profile curve representing the height variation as a function of position for the clay briquette with 10% ZPTs fired at 900 °C (Figure 13) shows marked fluctuations and significant irregularities, ranging from −60 μm to approximately 80 μm. The peak and valley values are more pronounced than those observed in the clay briquette alone, suggesting that adding 10% ZPTs increases the surface roughness.

The 3D image of the surface of the clay brick with 10% ZPTs (Figure 14) confirms the irregularities observed in the previous curve. The specific roughness values processed by 'cyber TECHNOLOGIES scan 8' software version 8.5 for this briquette are as follows: Ra = 6.48 μm, Rq = 8.55 μm, and Rz = 26.56 μm (rectangular raster) and Ra = 4.95 μm, Rq = 6.36 μm, and Rz = 17.84 μm (linear profile). These values show that the surface of the clay briquette with 30% ZPTs has a higher roughness than that of the clay briquette alone (Ra = 4.42 μm, Rq = 5.525 μm, and Rz = 17.68 μm). The 3D image highlights significant height variations with marked peaks and valleys, confirming that adding 10% ZPTs results in an increase in surface roughness. Compared to the briquette made with 100% clay, adding ZPTs appears to significantly alter the surface texture, making the briquette less homogeneous and rougher.

4.2. Physic-Mechanical Tests

4.2.1. Bulk Density of Dry Mixtures

The mass and volume of the mixtures prepared for briquetting reveal interesting trends in density. The mass of the briquettes gradually increases with an increase in the added percentage of ZPTs, showing a trend toward a greater material compactness with a greater amount of added ZPTs. This increase in mass is also reflected in the density, which increases linearly with the percentage of ZPTs. As shown in Figure 15, the bulk density of the samples without ZPTs is above 1.03 g/cm^3. The addition of ZPTs increases the bulk density of mixtures up to 1.37 g/cm^3, which is due to the relatively high unit weight of ZPTs. These results suggest that adding ZPTs to clay leads to denser briquettes, which may impact their strength and ability to withstand loads and stresses during use.

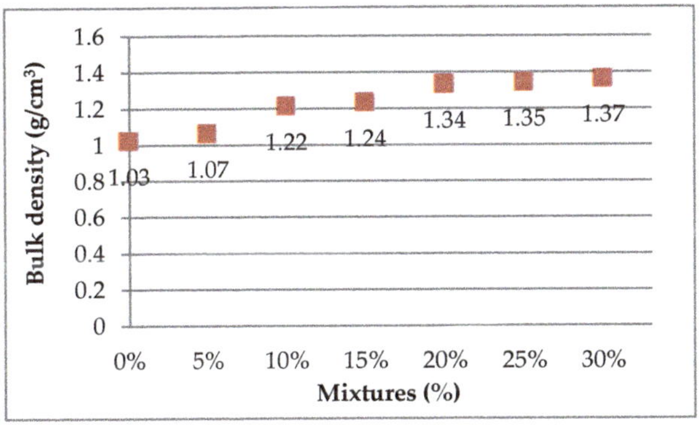

Figure 15. Variation in bulk density of different clay mixtures.

4.2.2. Plasticity Limits

The plasticity limit results of the mixtures, illustrated in Figure 16, show a clear inverse correlation between the addition of ZPTs and plasticity. Specifically, the plasticity limit decreases from 25.7% for pure clay to 17.09% at 30% ZPTs, indicating a significant reduction in the workability of the clay paste. This trend suggests that ZPTs acts as a de-plasticiser, reducing the clay matrix's cohesion and water retention capacity.

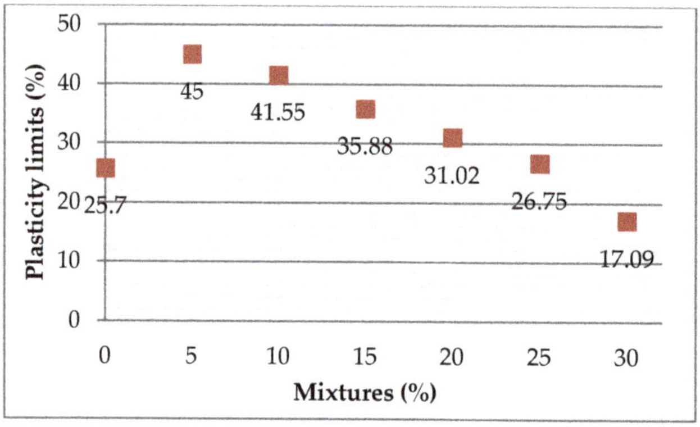

Figure 16. Results of plasticity limits.

Higher ZPT percentages may require additional processing adjustments, such as an increased water content or the use of plasticising additives, to maintain the necessary shaping properties. Based on the observed plasticity values, formulations with lower ZPT percentages (5%–15%) may offer the best compromise, maintaining a sufficient plasticity for shaping while benefiting from the influence of ZPTs in modifying the final ceramic properties.

4.2.3. Water Absorption

Figure 17 represents the results of water absorption tests in the following two cases: 900 °C and 1000 °C. At 900 °C, a general increase in water absorption is clear, suggesting an increased porosity of the briquettes. At 1000 °C, variations in water absorption are less pronounced.

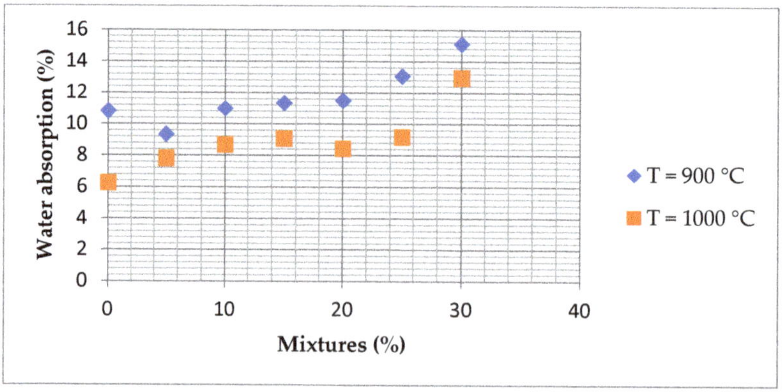

Figure 17. Water absorption as a function of mixtures fired at 900 °C and 1000 °C.

The increased water absorption observed in bricks containing ZPTs can be primarily attributed to these residues' high loss on ignition (LOI). The elevated LOI indicates a significant volatile matter content, leading to substantial gas release during firing. This gas evolution creates additional porosity within the ceramic matrix, increasing the capacity of the bricks to absorb water.

Despite the presence of fluxing agents (Ca and Mg), which generally promote sintering and densification, the dominant effect of gas release disrupts matrix consolidation, resulting in a porous structure.

4.2.4. Bulk Density of Fired Bricks

The bulk densities of the produced bricks with different proportions of ZPTs fired at different temperatures (900–1000 °C) are presented in Figure 18. The results reveal significant variations in the density depending on the added quantity of ZPTs and the firing temperature. As presented, the bulk densities of the samples without ZPTs, which were fired at 900 °C and 1000 °C, are above 1.18 and 1.40 g/cm^3, respectively. The addition of the ZPTs influences the bulk density of the bricks. At 900 °C, the density varies irregularly, with peaks observed at 5% and 10% of added ZPTs. On the other hand, at 1000 °C, ZPTs' density augmentation increases with an increase in ZPTs, reaching maximum values at 15% and 20% additions.

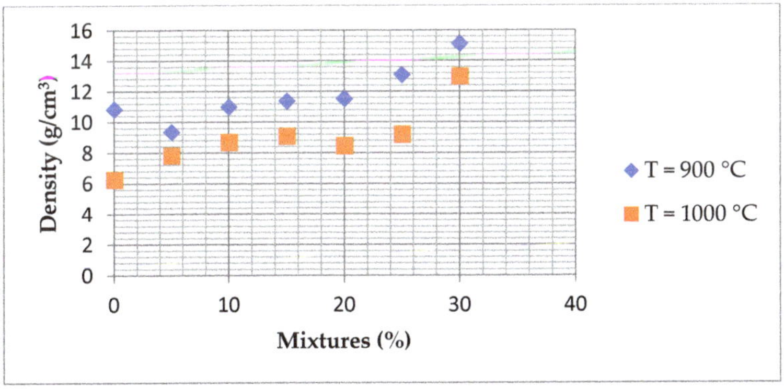

Figure 18. Graphical representation of density variation as function of mixtures and firing temperature.

The variation in bulk density reflects the interplay between densification and gas release during firing. At lower ZPT contents, fluxing agents (Ca and Mg) facilitate partial sintering, improving densification and leading to an initial increase in bulk density. However, as the ZPT proportion rises, the high loss on ignition (LOI) causes substantial gas evolution during thermal decomposition, generating an internal porosity that counteracts the densification process. This effect is more pronounced at 900 °C, where sintering remains incomplete, resulting in lower density values. In contrast, at 1000 °C, the densification effect becomes more pronounced due to enhanced sintering and vitrification promoted by CaO and MgO from ZPTs. This leads to a more compact structure, counteracting the porosity increase and resulting in a gradual bulk density increase of up to 15%–20% ZPT addition.

4.2.5. Linear Shrinkage

In Figure 19, a general reduction trend in linear shrinkage is observed with an increasing ZPT addition, suggesting less contraction of the briquettes at higher concentrations. This can be attributed to a decrease in clay minerals, particularly kaolinite, primarily responsible for shrinkage during drying. The reduction in these minerals limits the extent of water loss-induced contraction. Additionally, better adhesion and a more homogeneous structure of the materials at these concentrations reduce the internal stresses responsible for shrinkage. These findings highlight the influence of ZPT content on the physical properties of the briquettes, particularly their drying behaviour, which may affect their final dimensions and compatibility with intended applications.

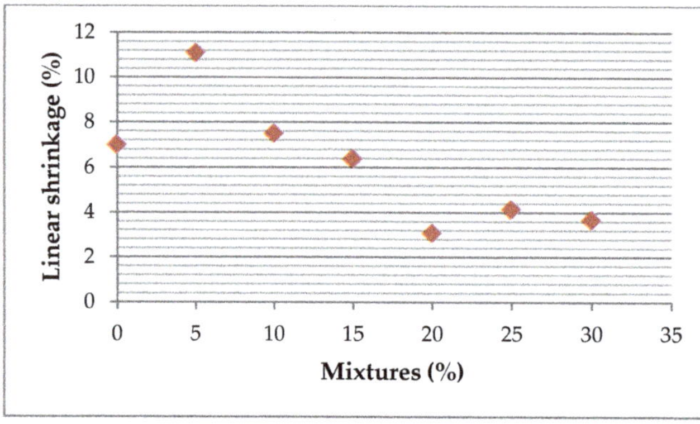

Figure 19. Linear shrinkage of bricks after drying.

Figure 20 presents the linear shrinkage results of bricks fired at 900 °C and 1000 °C as a function of ZPT content. The data indicate a general decreasing trend in shrinkage with an increasing ZPT percentage at both firing temperatures. At 900 °C, shrinkage starts at approximately 7% for the control sample (0% ZPTs) and gradually decreases with an increasing ZPT content. Similarly, at 1000 °C, shrinkage follows a comparable decreasing trend, with slightly higher values than 900 °C. This behaviour can be attributed to reducing clay minerals, particularly kaolinite, which limits contraction during drying and firing. The presence of ZPTs also influences the sintering process, potentially modifying densification mechanisms and reducing the internal stresses responsible for shrinkage.

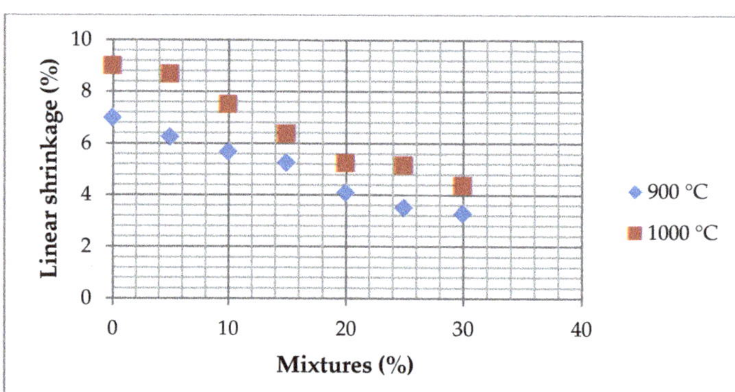

Figure 20. Linear shrinkage of bricks fired at 900 and 1000 °C with 0%–30% of added ZPTs.

4.2.6. Mechanical Resistance

The mechanical resistance tests conducted on fired clay bricks incorporating from 0 to 30 wt% of ZPTs at firing temperatures of 900 °C and 1000 °C, as presented in Figure 21, reveal significant variations in mechanical performance, strongly influenced by the mineralogical and chemical composition of ZPTs.

At 900 °C, the flexural strength (FS) decreases with increasing of ZPTs content. This decline suggests a disruption in the formation of a cohesive ceramic matrix, likely due to the presence of volatile compounds and organic residues in ZPTs, which decompose during firing. The high loss on ignition (LOI) of ZPTs leads to the release of gases (such as

CO_2 from carbonates and SO_2 from sulphates), increasing porosity and reducing the ability of the bricks to withstand flexural stresses. Additionally, the lower sintering activity of ZPT-containing mixtures at this temperature may result in weaker particle bonding.

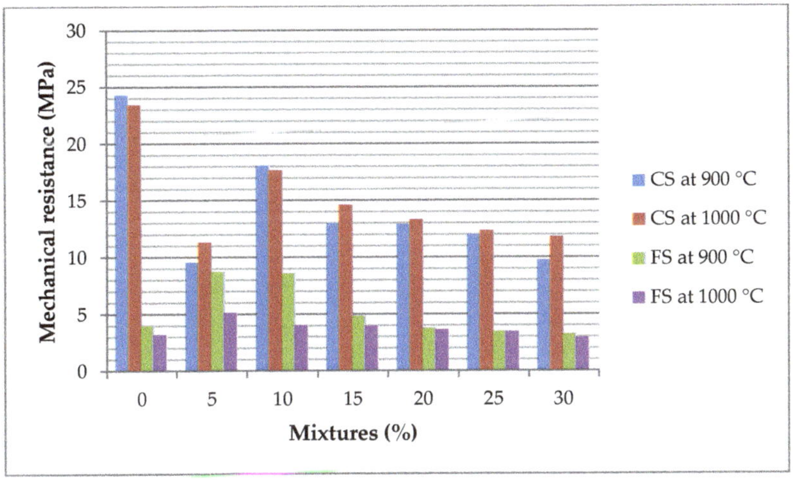

Figure 21. Mechanical resistance of fired clay bricks at 900–1000 °C with from 0 to 30 wt% of ZPT additions (CS: compressive strength and FS: flexural strength).

In contrast, the compressive strength (CS) at 900 °C does not follow a linear trend, but reaches a peak at 10% ZPTs before declining at higher concentrations. This behaviour can be attributed to the presence of iron oxides (Fe_2O_3) and calcium compounds (CaO, $CaCO_3$) in ZPTs, which can act as fluxing agents, promoting partial vitrification and enhancing densification at moderate ZPT levels. However, beyond 10% ZPTs, the increase in porosity due to excessive gas release counteracts these benefits, leading to a progressive decrease in compressive strength.

At 1000 °C, the flexural strength continues to decrease, confirming the negative effect of ZPT addition on the ability of the bricks to withstand bending stresses. The persistence of this trend suggests that, even at higher temperatures, the porosity induced by ZPTs prevents the formation of a dense and mechanically resistant microstructure. Meanwhile, the compressive strength exhibits more pronounced fluctuations, with peaks at 10% and 15% ZPTs. This non-linear behaviour is likely due to phase transformations occurring at high temperatures. At 1000 °C, iron oxides may contribute to the formation of liquid phases that enhance densification at intermediate ZPT levels. Still, excessive amounts of ZPTs could form fragile crystalline phases (such as excessive hematite Fe_2O_3 or anorthite $CaAl_2Si_2O_8$), which may compromise mechanical integrity.

These findings highlight the crucial roles of ZPT content and firing temperature in determining the mechanical properties of bricks. While small additions of ZPTs (up to 10%–15%) may improve compressive strength due to densification effects, excessive incorporation leads to an increased porosity and the potential formation of brittle phases, ultimately limiting the mechanical performance of the bricks. Therefore, optimising the mixture design requires a careful balance between ZPT content and firing conditions to achieve the desired mechanical properties for construction applications.

5. Conclusions

This study explored the feasibility of incorporating zinc processing tailings (ZPTs) as a partial replacement for clay in fired brick production. The experimental findings demonstrated that ZPTs affect the plasticity, sintering behaviour, and final properties of bricks, leading to challenges and opportunities in optimising formulations.

The plasticity tests showed a progressive decrease in plasticity with an increasing ZPT content, confirming its role as a de-plasticiser. This reduction in workability can affect the shaping and forming process, necessitating adjustments such as an increased water content or the use of plasticising agents.

A key finding was the influence of ZPTs' high loss on ignition (LOI = 35.46%), which induces gas release during firing. Despite the presence of fluxing agents (Ca and Mg) that generally promote sintering and densification, the dominant effect of volatile release led to an increased porosity, as evidenced by the higher water absorption values observed at 900 °C. The microstructural impact of this gas release resulted in a lower bulk density, indicating a need to optimise firing parameters to counterbalance porosity.

In terms of mechanical performance, an increasing ZPT content weakened compressive strength due to a higher porosity. However, at 10% ZPTs and 900 °C, a balance was achieved between mechanical integrity and durability, making this formulation the most suitable for practical application in brick manufacturing.

Compared to previous research on industrial waste incorporation in ceramics, this study provides a new perspective on the trade-off between sintering enhancement and porosity increase due to gas evolution. It highlights the need for process control strategies to mitigate the adverse effects of ZP41T decomposition. Future research should focus on optimising the thermal profile, exploring stabilising additives, or implementing alternative firing atmospheres to improve the densification and strength of ZPT-containing bricks.

Overall, this study advances the field by demonstrating that ZPTs can be effectively reused in brick production when carefully optimised, contributing to both sustainable waste management and innovative material development in the ceramics industry.

6. Prospects

Therefore, further in-depth research is essential to develop ZPT utilisation as a partial replacement for clay in fired bricks, with a focus on the following aspects:

- The thermal characteristics of ZPTs should be studied;
- The effect of ZPTs on the performance of fired bricks as a building material needs to be further explored;
- Heavy metal leaching in bricks made from ZPTs could be considered to precisely quantify the mobility of heavy metals after firing and to ensure the security and workability of ZPT bricks.

Author Contributions: Conceptualisation, A.M., B.F.P., V.N. and A.B.; methodology, A.M.; formal analysis, A.M., J.M.M.-A., L.B., I.M. and M.T.; investigation, J.M.M.-A., L.B., I.M. and M.T.; resources, A.M., B.F.P. and A.B.; writing—original draft preparation, A.M. and V.N.; writing—review and editing, A.M., J.M.M.-A., V.N. and M.T.; All authors have read and agreed to the published version of the manuscript.

Funding: The research presented in this paper had the financial support of the Ministry of Science, Technological Development and Innovation of the Republic of Serbia, within the funding of the scientific research work at the University of Belgrade, Technical Faculty in Bor, according to the contract with registration number 451-03-137/2025-03/200131.

Data Availability Statement: Data are contained within the article.

Conflicts of Interest: The authors declare no conflicts of interest.

References

1. Sun, R.; Gao, Y.; Yang, Y. Leaching of heavy metals from lead-zinc mine tailings and the subsequent migration and transformation characteristics in paddy soil. *Chemosphere* **2022**, *291*, 132792. [CrossRef]
2. Li, C.; Wen, Q.; Hong, M.; Liang, Z.; Zhuang, Z.; Yu, Y. Heavy metals leaching in bricks made from lead and zinc mine tailings with varied chemical components. *Constr. Build. Mater.* **2017**, *134*, 443–451. [CrossRef]
3. Ngayakamo, B.H.; Bello, A.; Onwualu, A.P. Development of eco-friendly fired clay bricks incorporated with granite and eggshell wastes. *Environ. Chall.* **2020**, *1*, 100006. [CrossRef]
4. Zhou, Y.; Duan, X.; Chen, T.; Yan, B.; Li, L. Mechanical properties and toxicity risks of lead-zinc sulfide tailing-based construction materials. *Materials* **2021**, *14*, 29. [CrossRef] [PubMed]
5. Nassiri, O.; Rhoujjati, A.; Moreno-Jimenez, E.; Hachimi, M.L.E. Environmental and geochemical characteristics of heavy metals in soils around the former mining area of zeïda (high moulouya, Morocco). *Water Air Soil Pollut.* **2023**, *234*, 110. [CrossRef]
6. Wiemes, L.; Pawlowsky, U.; Mymrin, V. Incorporation of industrial wastes as raw materials in brick's formulation. *J. Clean. Prod.* **2017**, *142*, 69–77. [CrossRef]
7. Behera, S.K.; Ghosh, C.N.; Mishra, K.; Mishra, D.P.; Singh, P.; Mandal, P.K.; Buragohain, J.; Sethi, M.K. Utilisation of lead-zinc mill tailings and slag as paste backfill materials. *Environ. Earth Sci.* **2020**, *79*, 1–18. [CrossRef]
8. Doğan-Sağlamtimur, N.; Bilgil, A.; Szechyńska-Hebda, M.; Parzych, S.; Hebda, M. Eco-friendly fired brick produced from industrial ash and natural clay: A study of waste reuse. *Materials* **2021**, *14*, 877. [CrossRef]
9. Li, R.; Yin, Z.; Lin, H. Research status and prospects for the utilisation of lead–zinc tailings as building materials. *Buildings* **2023**, *13*, 150. [CrossRef]
10. Yaras, A.; Sutcu, M.; Erdogmus, E.; Gencel, O. Recycling and immobilisation of zinc extraction residue in clay-based brick manufacturing. *J. Build. Eng.* **2021**, *41*, 102421. [CrossRef]
11. Lin, H.; Li, R.; Li, S. Fabrication of Lead–Zinc Tailings Sintered Brick and Its Effect Factors Based on an Orthogonal Experiment. *Materials* **2024**, *17*, 2352. [CrossRef] [PubMed]
12. Wang, G.; Ning, X.A.; Lu, X.; Lai, X.; Cai, H.; Liu, Y.; Zhang, T. Effect of sintering temperature on mineral composition and heavy metals mobility in tailings bricks. *Waste Manag.* **2019**, *93*, 112–121. [CrossRef]
13. Huang, Y.Y.; Qi, Y.Y.; Deng, L. Effect of ZnO on the Mineral Composition of Portland cement clinker. *Bull. Chin. Ceram. Soc.* **2017**, *36*, 1567–1572.
14. Nouairi, J.; Hajjaji, W.; Costa, C.S.; Senff, L.; Patinha, C.; da Silva, E.F.; Labrincha, J.A.; Rocha, F.; Medhioub, M. Study of Zn-Pb ore tailings and their potential in cement technology. *J. Afr. Earth Sci.* **2018**, *139*, 165–172. [CrossRef]
15. Luo, Z.; Tang, C.; Mu, Y.; Liu, X. Recycle of lead-zinc tailings in blended cement: Mechanical property and stabilisation/solidification of heavy metals. *Prepr. 2022 Miner.* **2019**, *9*, 710. [CrossRef]
16. Zhao, T.; Zhang, S.; Yang, H.; Ni, W.; Li, J.; Zhang, G.; Teng, G. Influence on fine lead-zinc tailings solidified/stabilised by clinker-free slag-based binder. *J. Environ. Chem. Eng.* **2022**, *10*, 108692. [CrossRef]
17. Gou, M.; Zhou, L.; Then, N.W.Y. Utilisation of tailings in cement and concrete: A review. *Sci. Eng. Compos. Mater.* **2019**, *26*, 449–464. [CrossRef]
18. Saedi, A.; Jamshidi-Zanjani, A.; Darban, A.K.; Mohseni, M.; Nejati, H. Utilization of lead–zinc mine tailings as cement substitutes in concrete construction: Effect of sulfide content. *J. Build. Eng.* **2022**, *57*, 104865. [CrossRef]
19. Saedi, A.; Jamshidi-Zanjani, A.; Mohseni, M.; Darban, A.K.; Nejati, H. Mechanical activation of lead–zinc mine tailings as a substitution for cement in concrete construction. *Constr. Build. Mater.* **2023**, *364*, 129973. [CrossRef]
20. Chen, Z.; Xiao, L.; Tao, Q.; Xie, L. Study on preparation and shielding effect of lead-zinc tailings sand mortar. *IOP Conf. Ser. Earth Environ. Sci.* **2019**, *300*, 022152. [CrossRef]
21. Wang, H.; Ju, C.; Zhou, M.; Dong, Y.; Hou, H.; Liu, S. Grinding kinetics of lead–zinc tailing powders and its optimal particle size as a pozzolanic admixture in cement mortar. *Adv. Powder Technol.* **2022**, *33*, 103730. [CrossRef]
22. Qu, G.; Ge, R.; Li, J.; Zhang, Y. Lead-Zinc Tailings-Based Construction Mortars/Preparation, Performance and Stabilization Mechanism of Heavy Metals. *Prep. Perform. Stab. Mech. Heavy Met.* **2022**. [CrossRef]
23. Peng, H.; Wang, D.; Sofi, M.; Mendis, P.; Zhou, Z.; Liu, J. Feasibility of Using Lead–Zinc Tailings to Produce Environmentally Friendly Ceramisite. *J. Mater. Civ. Eng.* **2021**, *402*, 331298. [CrossRef]
24. Li, C.; Zhang, P.; Li, D. Study on low-cost preparation of glass–ceramic from municipal solid waste incineration (MSWI) fly ash and lead-zinc tailings. *Constr. Build. Mater.* **2022**, *356*, 129231. [CrossRef]
25. Zhang, X.; Li, L.; Hassan, Q.U.; Pan, D.; Zhu, G. Preparation and characterisation of glass ceramics synthesised from lead slag and lead-zinc tailings. *Ceram. Int.* **2023**, *49*, 16164–16173. [CrossRef]
26. Luo, Z.; Guo, J.; Liu, X.; Mu, Y.; Zhang, M.; Zhang, M.; Tian, C.; Ou, J.; Mi, J. Preparation of ceramsite from lead-zinc tailings and coal gangue: Physical properties and solidification of heavy metals. *Constr. Build. Mater.* **2023**, *368*, 130426. [CrossRef]
27. Akkaya, U.G.; Cinku, K.; Yilmaz, E. Characterization of strength and quality of cemented mine backfill made up of lead-zinc processing tailings. *Front. Mater.* **2021**, *8*, 740116. [CrossRef]

28. Zhang, F.; Li, Y.; Zhang, J.; Gui, X.; Zhu, X.; Zhao, C. Effects of slag-based cementitious material on the mechanical behavior and heavy metal immobilisation of mine tailings based cemented paste backfill. *Heliyon* **2022**, *8*, e10695. [CrossRef]
29. Feng, Q.M.; Wang, W.Q.; Zhang, B.L.; Huang, Y. Research on technics of lightweight baking-free brick made of lead-zinc ore tailings from Qinghai Province. *Non-Met. Mines* **2011**, *34*, 6–8.
30. Li, C.; Xu, Y.L.; Yu, Y.; Lin, Y.B.; Lin, J. The preparation and research of unburned and absorptive bricks of Pb-Zn mine tailings. *Mater. Sci. Technol.* **2016**, *24*, 46–51.
31. Lograda, T.; Harkati, Z.; Adel, K.; Ramdani, M. Heavy metals accumulation in species from mine Karzet Youcef (Algeria). *Word. J. Pharmaceut. Res.* **2016**, *5*, 250–260.
32. Belguidoum, A.; Lograda, T.; Ramdani, M. Heavy metals accumulation in Hertia cheirifolia along the highway in Setif region, Algeria. *Biodiversitas J. Biol. Divers.* **2020**, *21*. [CrossRef]
33. Omara, R. Ecological and Geochemical Assessment of the Environment in the Zinc Ore Recovery Zone; Case of CHAABET EL-HAMRA Mining Complex (Algeria). In *Advances in Green Energies and Materials Technology: Selected Articles from the Algerian Symposium on Renewable Energy and Materials (ASREM-2020)*; Springer: Singapore, 2021.
34. *ASTM D4318-17e1*; Standard Test Methods for Liquid Limit, Plastic Limit, and Plasticity Index of Soils. ASTM International: West Concord, PA, USA, 2017.
35. *ASTM D4943-18*; Standard Test Method for Shrinkage Factors of Soils by the Wax Method. ASTM International: West Concord, PA, USA, 2018.
36. *ASTM C29/C29M-17a*; Standard Test Method for Bulk Density ("Unit Weight") and Voids in Aggregate. ASTM International: West Concord, PA, USA, 2017.
37. *ASTM C373-18*; Standard Test Method for Water Absorption, Bulk Density, Apparent Porosity, and Apparent Specific Gravity of Fired White ware Products. ASTM International: West Concord, PA, USA, 2018.
38. *ASTM C326-09*; Standard Test Method for Drying and Firing Shrinkages of Ceramic White ware Clays. ASTM International: West Concord, PA, USA, 2016.
39. *ASTM C67-22*; Standard Test Methods for Sampling and Testing Brick and Structural Clay Tile. ASTM International: West Concord, PA, USA, 2022.
40. *ASTM C674-88*; Standard Test Method for Flexural Properties of Ceramic White ware Materials. ASTM International: West Concord, PA, USA, 2019.
41. Wang, S.; Gainey, L.; Baxter, D.; Wang, X.; Mackinnon, I.D.; Xi, Y. Thermal behaviours of clay mixtures during brick firing: A combined study of in-situ XRD, TGA and thermal dilatometry. *Constr. Build. Mater.* **2021**, *299*, 124319. [CrossRef]
42. Milićević, S.; Milošević, V.; Povrenović, D.; Stojanović, J.; Martinović, S.; Babić, B. Removal of Heavy Metals from Aqueous Solution Using Natural and Fe(III) Oxyhydroxide Clinoptilolite. *Clays Clay Miner.* **2013**, *61*, 508–517. [CrossRef]

minerals

MDPI

Article

Optimization of Magnetization Roasting and Magnetic Separation for the Recovery of Iron from Low-Grade Iron Plant Tailings

Anele Shamase * and Willie Nheta

Mineral Processing and Technology Research Centre, Department of Metallurgy, School of Mining, Metallurgy and Chemical Engineering, Faculty of Engineering and The Built Environment, University of Johannesburg, Johannesburg, Doornfontein P.O. Box 17911, South Africa; wnheta@uj.ac.za

* Correspondence: ashamase@uj.ac.za

Abstract: This study employs response surface methodology (RSM) with a custom optimal design to develop and optimize iron (Fe) grade and recovery through magnetization roasting followed by induced dry-roll magnetic separation. The relationships between the independent and dependent variables are investigated. The effect of roasting temperature, magnetization roasting time, magnetic field intensity, rotor speed, and product splitter position for the induced dry-roll magnetic separator on Fe grade and recovery are studied. Suitable models are generated to predict the optimum operating conditions. An analysis of Variance (ANOVA) is employed to validate the developed regression models' adequacy and assess the main and interaction-related effects on Fe grade and recovery. During magnetization roasting, a satisfactory Fe grade of 66.8% with a recovery of 16.7% was obtained under optimal conditions of 1050 °C for 97 min. Conversely, after the induced dry-roll magnetic separator optimization, an Fe grade of 66.1% with a recovery of 60.2% was achieved under optimum conditions of 0.105 T for magnetic field intensity, 70 Hz for rotor speed, and an 11 mm product splitter position. This study effectively illustrates how RSM can model the processes of magnetization roasting and induced dry-roll magnetic separation, particularly concerning the operating parameters used for treating iron ore plant tailings. Furthermore, it highlights the efficiency of this methodology in generating substantial insights in a short timeframe while minimizing the number of experiments conducted.

Keywords: response surface methodology; modeling; low-grade iron plant tailings; magnetization roasting; induced dry-roll magnetic separation

1. Introduction

Iron is a vital metal that is crucial in advancing modern societies [1]. It constitutes approximately 95% of all metals utilized annually across various sectors, including in structural engineering, maritime, automotive, and general industrial applications [2]. The continuous rise in steel demand has led to a marked increase in the extraction of crude iron, resulting in the gradual depletion of high-grade iron ore reserves [3]. Consequently, there is a pressing need to process iron from secondary sources such as low-grade iron ores, fines/slimes, and tailings to satisfy the growing requirement for high-quality iron in steel production. The extraction of iron ore generates substantial amounts of tailings and fines, which accumulate in tailing dams and pose various environmental challenges [3]. The average iron content in these tailings is estimated to be around 40%, and, over extended periods of production, this can lead to excessive residue accumulation. This issue is particularly significant in operations that target lower-grade iron ores. According to reports by Dauce et al. (2019) [4] and Yang et al. (2018) [5], the annual generation of iron ore tail (IOT) ranges from approximately 632 million tonnes to 1.4 billion tonnes, highlighting its potential as a secondary source of iron to meet ever-increasing market demand.

Despite this potential value, few commercial operations aim to extract iron from tailings for industrial use. This is mainly due to the complex mineralogy of the tailings, which contain a high amount of gangue material and present challenges in liberating the iron minerals [6]. Traditional beneficiation technologies, such as wet magnetic separation, leaching, microwave roasting, sulfidation roasting, sodium carbonate (Na_2CO_3) roasting, flotation, and some gravity concentrations such as shaking table, jig, reflux classifier, teetered bed separator, multi-gravity separator, and spiral demonstrate some limitations in recovering iron from difficult-to-treat iron-bearing secondary sources.

Although several investigations on the recovery of iron from low-grade iron ore were also achieved using flotation through reverse anionic quartz flotation, magnetic separation remains one of the most effective separation methods to beneficiate iron. However, it has been found that most low-grade iron ores typically contain considerable amounts of hematite occurring together with siderite and limonite, which are characterized by a paramagnetic behavior, making them a refractory and challenging to treat using magnetic separation [7]. Recently, a new pyrometallurgical pretreatment method, based on the chemical transformation of paramagnetic Fe into ferromagnetic minerals such as magnetite (Fe_3O_4) or maghemite (γ-Fe_2O_3), suitable for magnetic separation using pyrometallurgical processing (e.g., magnetizing roasting) followed by conventional magnetic separation, has been developed [7,8]. Several studies have demonstrated that the transformation of iron ores under thermal pretreatment can be enhanced during free moisture removal at temperatures ranging from 120 to 140 °C, followed by the dehydroxylation of limonite when the temperature ranges between 250 °C and 350 °C. Further increases in temperature between 490 °C and 630 °C lead to siderite decomposition, while hematite's conversion into magnetite occurs at around 850 °C [9,10].

Magnetization roasting has emerged as an attractive area of research and is becoming a promising technology for producing high-quality iron concentrates from low-grade ores and secondary iron sources [6,9,11–18]. The process consists of a partial reductive roasting process that transforms hematite iron oxide with a high oxidative state (Fe^{3+}) into magnetite, a ferrimagnetic compound with a lower oxidative state (Fe^{2+}). This reduction occurs under oxygenated conditions in the presence of carbon, which, at high temperatures, leads to CO_2 generation due to the oxygen (O_2) contained in the air being responsible for the boudoir reaction occurrence and producing the CO promoting an exothermic reaction where the ratio CO/CO_2 must be carefully regulated to monitor the reduction of hematite into magnetite [19]. After magnetization roasting, the resulting product is subjected to low-intensity magnetic separation to effectively remove impurities such as alumina (Al_2O_3) and silica (SiO_2) from the roasted sample [20].

Different reductants are used during the reduction process; the reductants, amongst others, include coking coal and agro-industrial-biomass, such as straw type (corn straw, rice), sawdust, starch, cow dung, etc. The choice of using biomass as opposed to conventional coking coal is because of its advantageous characteristics, which means it falls under the renewable category and provides a green process [18]. The use of biomass addresses the current SO_2, NO_x, and CO_2 emission challenges by using coal as a reducing agent [21,22]. Additionally, biomass naturally has a high amount of volatile carbonaceous matter, which creates additional reducing gases such as methane (CH_4), hydrogen (H_2), carbon dioxide

(CO_2), and carbon monoxide (CO). Therefore, CO is not the only reducing agent, as in the case of coal, which helps in generating more heat and accelerates the reduction process [15].

The current study aims to determine the impact of various parameters on the recovery and grade of iron (Fe) during magnetization roasting, followed by an induced dry-roll magnetic separator. A design of experiment (DoE) methodology using response surface methodology (RSM) and an optimal custom design was employed to optimize the beneficiation process. Therefore, this was achieved by utilizing a systematic approach incorporating multidimensional combinations of input variables in an optimal response surface methodology design. This approach helps to identify the optimal working conditions for achieving satisfactory iron grade and recovery.

2. Materials and Methods

2.1. Materials

The reductants used in this study were macadamia nutshells, which were crushed and pulverized to a size that could pass through a 100 µm sieve. The specifications for the reducing agents were analyzed using proximate and ultimate analyses. The moisture, volatile matter, and ash content were analyzed using a muffle furnace, model AAF 11/3, manufactured by Carbolite Gero Ltd., Sheffield, United Kingdom. The calorific value was analyzed using a combustion calorimeter, model CAL3K-F, manufactured by DDS CALORIMETERS, Randburg, South Africa. The ultimate analyses were performed using a CHNS analyser, model flash 2000, manufactured by Thermofisher scientific, Sunnyvale, CA, USA. The results are detailed in Table 1.

Table 1. Analysis of macadamia nutshells.

CV (MJ/kg)	Proximate Analysis (%)				Ultimate Analysis (%)			
	Moisture	VM	FC	Ash	Sulfur (S)	Carbon (C)	Hydrogen (H)	Nitrogen (N)
18.565	8.71	74.90	14.20	2.18	0.767	48.29	5.43	0.35

CV—calorific value, VM—volatile matter, FC—fixed carbon.

About 100 kg of a representative sample was collected from the iron plant tailings dam located in the Kathu District of South Africa's Northern Cape province. This sample was then crushed in a closed-circuit crusher to achieve a particle size of 100% passing 2 mm, using a Laboratory Roll Crusher, model WT-1873-2, manufactured by Metso, Denver, CO, USA. After crushing, the sample was homogenized and prepared for subsequent characterization and optimization studies.

Particle size distribution analysis was conducted on a 5000 g sample using a vibratory sieve shaker with manual amplitude adjustment settings ranging from 0.5 to 3 mm, model Analysette 3 SPARTAN, manufactured by Fritsch, Idar-Oberstein, Germany. The sequence of wire mesh sieves used was determined following the U.S. Standard Scale ratio, based on the square root of 2. The chemical composition of the head sample was analyzed using an X-ray fluorescence (XRF) model ZSX Primus II, manufactured by Rigaku, Tokyo, Japan, equipped with EZScan software, version 3. The phase analysis was also performed on the head sample using an X-ray diffractometer (XRD), model Ultima IV, manufactured by Rigaku, Tokyo, Japan, to determine the mineralogical species equipped with PDXL analysis version 2 software loaded with PDF ICDD cards. The sample microstructure and morphology were observed under scanning electron microscopy (SEM), model Vega 3XMU version 3, manufactured by TESCAN, Kohoutovice, Czech Republic, associated with Energy Dispersive X-ray Spectroscopy (EDS) to determine the chemical composition of selected areas based on contrast differences to ascertain the mineralogical phase. Images of samples were collected using either a backscatter or a secondary electron detector working under nano-space at the working distance of 15 mm while exciting electrons in the tube at a high voltage of 20 KVA using the Vega 3 software.

2.2. Methods

The beneficiation of magnetite took place in two stages, consisting of roasting magnetization followed by magnetic separation. During the magnetization roasting experiments, a mixture of iron ore plant tailings and a reductant (macadamia nutshells) in a 1:4 ratio was loaded into an Atlan laboratory muffle furnace model 1979. It should be noted that, considering carbon as the significant reductant, the expected partial reduction reaction of hematite into magnetite is represented by Equation (1) [12]. The calcined samples generated at each condition were subjected to the induced dry-roll magnetic separator model I.M.R Roll 5′, manufactured by Eriez Manufacturing Co, Erie, PA, USA, operating at 0.15 T magnetic intensity to evaluate the efficiency of the roasting. The emanating magnetic separator products were assessed using XRF, and the performance of the magnetization roasting was evaluated based on the Fe_T recovered in the magnetic separator using Equation (2).

$$3Fe_2O_{3(s)} + CO_{(g)} \leftrightharpoons 2Fe_3O_{4(s)} + CO_{2(g)},\ T > 550\ ^{\circ}C \tag{1}$$

$$Fe_T recovery, \% = \frac{100{\bullet}C_{Fe}{\bullet}m_c}{f_{Fe}{\bullet}F_f} \tag{2}$$

where C_{Fe} and m_c are the total grade of iron in the non-magnetic fraction and mass of the non-magnetic fraction, respectively. f_{Fe} and F_f are the iron grade in the feed and mass of the feed, respectively.

After determining the conditions that lead to optimum magnetization roasting, the bulk sample was generated to enhance the efficiency of the induced dry-roll magnetic separation. Following each test run, the magnetic and non-magnetic fractions were analyzed using XRF to assess the effectiveness of the separation process. The flowsheet illustrating the experimental procedure is presented in Figure 1.

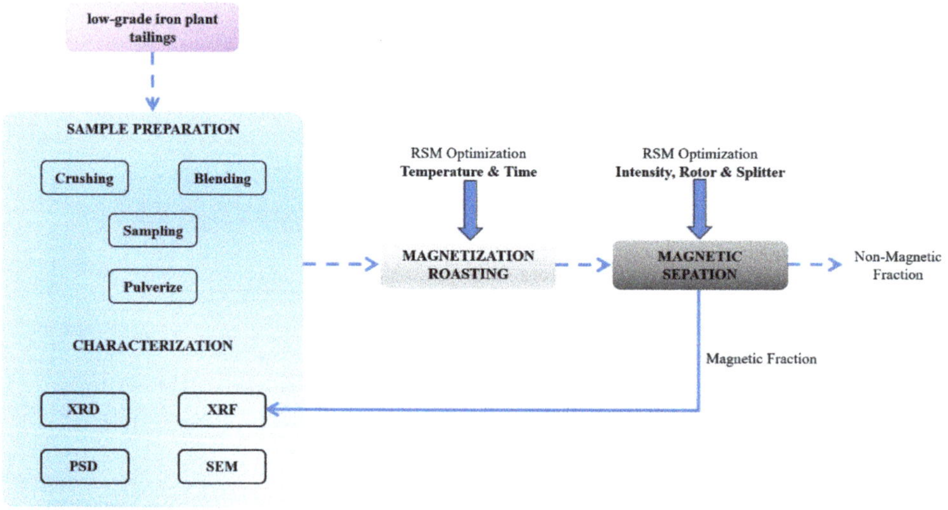

Figure 1. Experimental procedure flowsheet.

2.2.1. Magnetic Susceptibility Test

Prior to optimizing the magnetic separation, a magnetic susceptibility test was performed on the roasted bulk sample. This was performed to estimate the operating conditions, leading to the range in which the separation should occur. Therefore, after the attempt tests, the boundary levels of different parameters used in the DoE-RSM were defined based on the outcomes. Hence, in the current investigation, the roasted sample at an average temperature of 1000 °C for 30 min was subjected to the average magnetic field, rotor speed, and product splitter of 0.1 T, 20 Hz, and 10 mm, respectively.

2.2.2. Optimization of the Magnetic Separation Using the Design of Experiment (DoE) Through Response Surface Methodology (RSM)

The Design of Experiments (DoE) method is an optimization technique mainly used to determine and control the practical factors prompt to attain favorable conditions for a process under investigation. It is used to select independent variables (process parameters), significantly affecting the process development of the mathematical model fitting the obtained data and the process optimization [23]. This technique limits non-controllable factors by recognizing effective parameters based on the statistical approaches to the problem. Therefore, response surface methodology (RSM) is one of the experimental design tools mainly used to show the correlation between experimental variables (known as independent variables) and responses (known as dependent variables). As a result, it combines statistical methods for experimental modeling and process parameter optimization, where various independent variables influence the expected outcome [24].

In the current investigation, the optimal custom RSM design was utilized to examine the effect of the main parameters and their interactions on the Fe grade and Fe recovery of the concentrated product obtained from magnetization roasting and an induced dry-roll magnetic separator. This study employs Design-Expert® version 13.0.5.0 software. This software helps to investigate the significance of the responses in correlation with the variables used, using the regression and graphical analysis of experimental data on magnetite extraction during the two stages, "roasting, magnetization, and magnetic separation." The number of experiments for the optimization study was achieved using RSM through the optimal custom design. The trial sequence was randomized to reduce the impact of the uncontrolled elements.

Two roasting factors and three magnetic separation factors were chosen as the independent variables, while the Fe recovery and Fe grade were considered dependent variables (responses). A five-level two-factor system was designed using an optimal design configuration for magnetization roasting optimization experiments. The independent variables considered were roasting temperature (A) and roasting time (B). Each variable was coded at five levels (L1, L2, L3, L4, and L5), leading to 12 experimental runs, as shown in Table 2.

In the case of the optimization experiments of induced dry-roll magnetic separation, three key variables were considered: magnetic field intensity (A), rotor speed (B), and product splitter position (C). The magnetic field intensity and product splitter position were coded with three levels (L1, L2, and L3), while the rotor speed was coded with four levels (L1, L2, L3, and L4). A total of 20 experiments were designed using the optimal design configuration, as shown in Table 2.

The correct application of the method requires a multistep approach: defining the problem (process variables, responses, what can be measured, and what can be optimized), strategy selection (selection of the variables or process optimization), designing the experimental matrix, conducting experiments, analyzing the obtained data, developing the mathematical model and its evaluation, and validating the model [25].

Table 2. The level of variables for magnetization roasting optimization and the induced dry-roll magnetic separator.

Variable	Factors (Coded X_i)	Experimental Design (Coded Level)				
		L_1	L_2	L_3	L_4	L_5
Magnetization Roasting Optimization						
Roasting temperature (°C)	*A*	500	650	800	950	1100
Roasting time (min)	*B*	15	30	60	90	120
Induced dry-roll magnetic separator optimization						
Magnetic field intensity (T)	*A*	0.05	0.1	0.14	-	-
Rotor speed (Hz)	*B*	20	40	60	70	-
Product splitter position (mm)	*C*	5	10	15	-	-

The RSM technique defines a model for each independent variable that expresses the primary and interactive effects of the factors on each independent variable. Hence, Response Surface Methodology (RSM) assumes a mathematical relationship between the response variable and the factors involved. The relationship is expressed by a quadratic equation linearly, two-order linearly, or quadratically, as indicated by Equation (3), represented as a multivariate model [25,26].

$$y = \beta_0 + \sum_{i=1}^{k} \beta_i x_i + \sum_{i=1}^{k} \sum_{i=1}^{k} \beta_{ij} x_i x_j + \sum_{i=1}^{k} \beta_{ii} x_i^2 \quad (3)$$

where x_i is the factor and the β_i and β_j are the regression coefficients.

The viability of the optimal conditions for the roast magnetization and the upgrading of the magnetite processes were determined using the optimal model prediction represented as a linear factor's equation. After that, the analysis of variance (ANOVA) was analyzed, laying on the criteria such that the *p*-value of the coefficient under investigation was $\leq$0.05 and included their comparison with the statistical references to check the sufficiency of the proposed model, while the quality of fitting was assessed using the goodness of fit by checking R^2 and the adjusted and predicated R^2 value for the significance of the model. Furthermore, to correlate the quality of the outcomes, the predicted and actual values are presented and discussed [27]. However, the main and interactive effects were also visualized and explained in correlation with the ANOVA. At the same time, counter and surface plots were displayed and elaborated to understand whether the optimization consisted of maximization or minimization. Ultimately, the disability was used to determine optimum operating conditions [28].

3. Results

3.1. The As-Received Sample Characterization

The chemical composition of the sample, detailed in Table 3, reveals that the head sample contains 82.76% Fe_2O_3, corresponding to a total iron (Fe) of 57.1%. Although the iron concentration in the iron plant tailings is substantial, it cannot be utilized directly for economic purposes due to significant impurities. The sample contains approximately 4.1% Al_2O_3 (2.2% Al) and 10.9% SiO_2 (5.11% Si). The assigned objective was upgrading the material to achieve at least 65% Fe. Additionally, the combined average of Al_2O_3 and SiO_2 must not exceed 5% to satisfy the required market specifications.

Table 3. Chemical composition of the as-received sample.

Composition (%)														
Fe	Na	Mg	Al	Si	S	Cl	K	Pb	Ca	Ti	Cr	Mn	O	Ca
57.1	0.02	0.03	2.18	5.11	0.04	0.01	0.71	0.10	0.08	0.09	0.02	0.57	34.04	0.08

The analysis of the XRD spectrum for the iron tailings sample, as shown in Figure 2, revealed that the sample was dominated by the presence of Hematite occurring with traces of limonite characterized by paramagnetic behavior. Further examination indicated that Si, Al, and other elements were found in the form of quartz (SiO_2), sillimanite (Al_2SiO_5), and traces of feldspar (Na, K, Ca), Al (Al, Si)$_3$O$_8$. According to Srivastava et al. (2001) [12], quartz and sillimanite can result in a high slag volume, which may lead to the formation of viscous slag during the iron-making process. Therefore, it is crucial to reduce these impurities through beneficiation to enhance the productivity of iron-making operations.

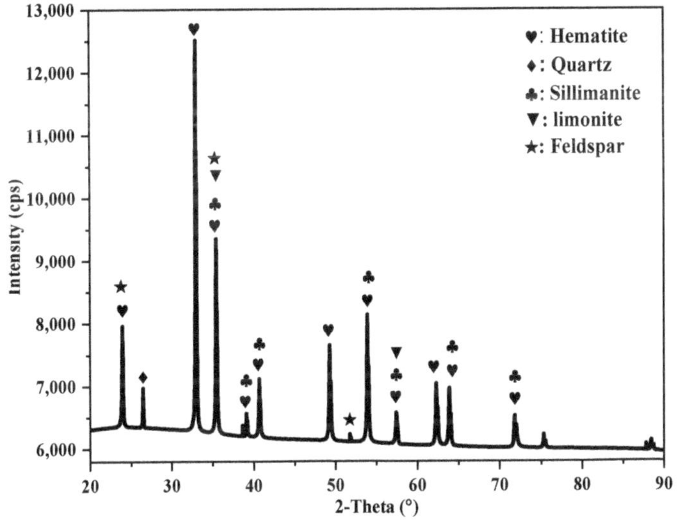

Figure 2. Mineralogical phases of the as-received sample.

The morphology of the as-received sample was analyzed using SEM, while the chemical composition of the selected areas was determined using EDS. The micrograph results are shown in Figure 3, and the EDS spectra analysis results are provided in Table 4. The micrograph indicates that the sample is characterized by two main contrasts, indicating that the hematite phase (light gray) is interlocked into the silicate one (dark gray) and vice versa. It should be noted that the occurrence of the magnetite makes separation challenging. Therefore, a fine milling process was required to liberate Fe phases.

The bright spots that indicate heavier compounds are identified as Fe_2O_3, whereas the gray spots, predominantly composed of lighter compounds, reveal the presence of minerals such as quartz and sillimanite. The EDS analysis confirmed that the sample primarily comprises hematite, sillimanite, and quartz, consistent with the XRD results. Furthermore, it was noted that silica is distributed throughout the hematite matrix. The circled areas in the mapping, as shown in Figure 4, confirm the evidence of the interlinked Al-, Fe-, K-, and Si-bearing minerals.

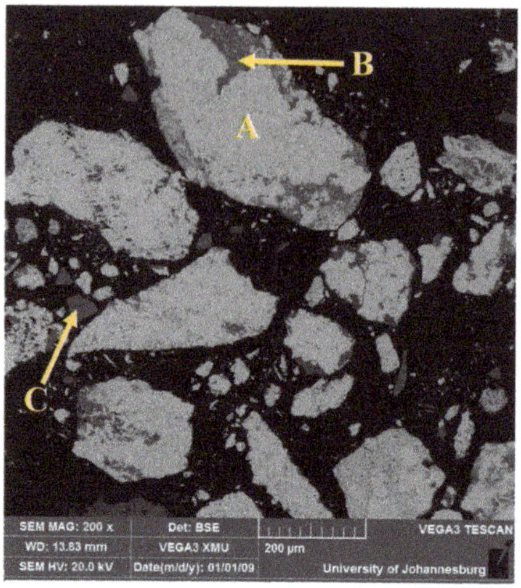

Figure 3. SEM image of the as-received sample for spectrum A, B and C.

Table 4. EDS result of the spectra of the as-received sample.

Spectrum	Composition (%)							Probable Phases
	O	Al	Si	K	Mn	Fe	Mg	
A	63.46	0.00	0.69	0.00	0.00	35.84	0.00	Hematite
B	73.95	7.98	11.38	0.78	0.00	4.72	0.00	Quartz, Sillimanite
C	62.09	13.36	17.09	3.54	0.13	3.34	0.13	Quartz, Sillimanite

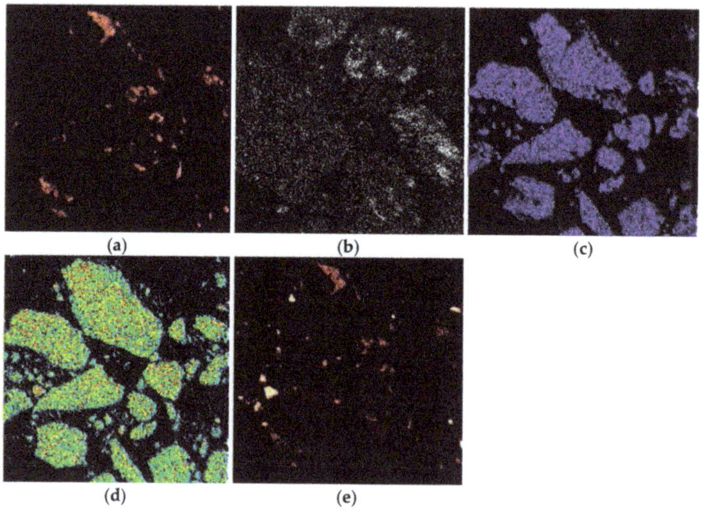

Figure 4. Mapping analysis of the as-received iron plant tailings samples showing element concentrations: (**a**) Al; (**b**) K; (**c**) Fe; (**d**) O; (**e**) Si.

The particle size distribution of the as-received iron plant tailings was analyzed using wire mesh sieves, and the results are illustrated in Figure 5. The size distribution curve exhibits a bimodal profile, with sections covering the size ranges of 40,000–1180 μm and 1180–75 μm, corresponding to 40% and 60% splits, respectively. This suggests that the iron plant tailing sample originates from a processing route that employs various beneficiation techniques with different size specifications, similar to those described by Das and Rath (2020) [29]. The analysis further indicated that the F_{80} of the iron tailing sample was 11,300 µm, encompassing a wide range of size fractions from 75 to 4000 µm, which points to a mixture of fine and coarse particles. Given that the primary objective of this project involves magnetic separation, it is crucial to compare the size distribution results of the as-received sample with the recommended particle size distribution for the effective operation of an induced dry-roll magnetic separator. This optimal range lies between 100 µm and 3000 µm, as Wills and Napier-Munn (2006) [30] specified. Therefore, the iron plant tailings must undergo comminution before processing to conform to these specifications.

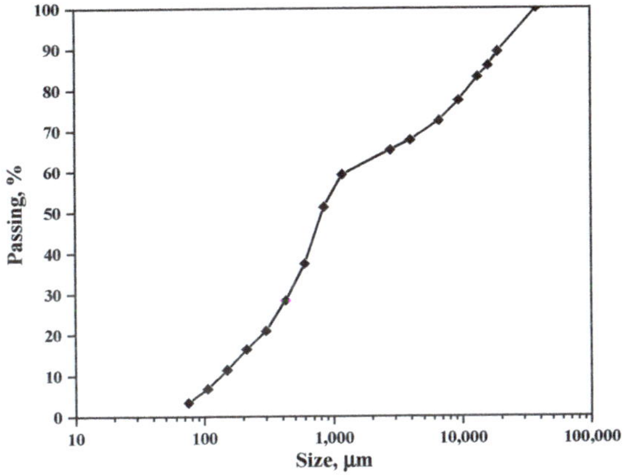

Figure 5. Particle size distribution of the head sample of iron plant tailings.

3.2. Magnetic Susceptibility Test Results

The magnetic susceptibility test was performed on the un-roasted and roasted samples to evaluate the achievement of the transformation by applying the magnetic separation and comparing the Fe grades. The results presented in Figure 6 reveal that the grade of Fe in the concentrate of the unroasted sample was 61.8%. However, the determination of the Fe grade on the magnetized sample subjected to the same conditions of magnetic separation was found to be 66.1% as a sign of the ferrimagnetic material formation.

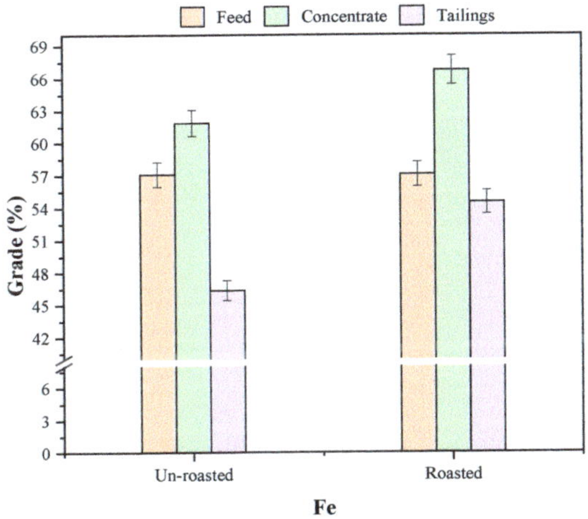

Figure 6. Magnetic susceptibility test results.

3.3. Optimization of the Magnetization Roasting Process Parameters

A two-factor five-level RSM optimal design was employed to assess the significance of roasting temperature (A) and time (B) and their interactions on the Fe grade and recovery achieved through magnetization roasting. A total of 12 experimental runs were proposed using Design-Expert® software version 13.0.5.0. Table 5 presents an overview of the experimental design and the observed responses for optimizing magnetization roasting.

Table 5. Experimental plan of optimal design with observed values for magnetization roasting optimization.

Run	Variables		Responses	
	A: Roasting Temperature (°C)	**B: Roasting Time (min)**	**Fe Grade (%)**	**Fe Recovery (%)**
1	500	15	61.91	11.47
2	1100	60	64.45	16.09
3	500	90	65.03	8.87
4	800	60	66.23	16.34
5	800	120	66.75	16.27
6	1100	15	65.92	16.27
7	650	30	62.11	15.9
8	1100	90	65.84	16.07
9	800	15	64.13	15.77
10	800	60	66.23	16.34
11	500	60	64.71	8.23
12	500	120	65.55	9.32

Two statistical models were developed for each response variable, i.e., Fe grade and Fe recovery. A linear first-order model was utilized to represent the relationship between Fe grade and each input factor derived from the experimental data [31]. Conversely, a quadratic second-order polynomial model described the relationship between controllable variables and Fe grade and recovery. The empirical coded equations for the Fe grade and Fe recovery are shown in Equations (4) and (5), respectively.

$$\text{Fe grade (\%)} = 65.19 + 0.92\,A + 1.41\,B \tag{4}$$

$$\text{Fe recovery (\%)} = 16.05 + 3.31\,A - 0.3\,B + 0.83\,A \times B - 3.46\,A^2 + 0.75\,B^2 \tag{5}$$

where the Fe grade and recovery are given in percentage, roasting temperature (A) is in °C, and roasting time (B) is in minutes.

The analysis of variance (ANOVA) was utilized at a 5% significant level to validate the adequacy of the developed regression models for Fe grade and recovery for a p-value ≤ 0.05. Tables 6 and 7 present the ANOVA analysis conducted for Fe grade and Fe recovery during the magnetization roasting optimization.

Table 6. ANOVA analysis for Fe grade in the magnetization roasting.

Source	Sum of Squares	DF	Mean Square	F-Value	*p*-Value	Remark
Model	15.32	2	7.66	5.96	0.0225	significant
A—Temperature	5.83	1	5.83	4.54	0.062	not significant
B—Time	11.32	1	11.32	8.81	0.0158	significant
Residual	11.56	9	1.28			
Cor Total	26.88	11				
Adeq precision: 6.7872						

Table 7. ANOVA analysis for Fe recovery in the magnetization roasting.

Source	Sum of Squares	DF	Mean Square	F-Value	*p*-Value	Remark
Model	118.46	5	23.69	23.64	0.0007	significant
A—Temperature	70.30	1	70.30	70.13	0.0002	significant
B—Time	0.4480	1	0.4480	0.4470	0.5286	not significant
$A \times B$	1.91	1	1.91	1.91	0.2163	not significant
A^2	30.84	1	30.84	30.77	0.0015	significant
B^2	1.19	1	1.19	1.19	0.3175	not significant
Residual	6.01	6	1.00			
Cor Total	124.47	11				
Adeq precision: 11.6						

Table 6 shows that a fissure test value (F-value) of 5.96 implies that the Fe grade model is significant. There is only a 2.25% chance that an F-value this large could occur due to noise.

The value of Adequate Precision (Adeq Precision) serves as a signal-to-noise ratio to evaluate the sufficiency of a model. Two hypotheses were formulated to assess the sufficiency [32].

- The "null hypotheses—H_0" suggests that the "Adeq Precision" value is >4: the predicted model fits the data and indicates adequate model discrimination.
- The "alternate hypotheses—H_a" implies that the value of "Adeq Precision" is <4, suggesting that the predicted model does not adequately fit the data and exhibits insufficient model discrimination.

With an "Adeq Precision" ANOVA value of 6.7872, which exceeds 4, we can conclude that the model is sufficient and can be reliably used to navigate the design space [33].

The values of *p*-values below 0.05 indicate that the model terms are statistically significant. In the context of Fe grade, the results reveal that term B is significant. The F-value assesses the main effects and their interactions, with larger F-values indicating greater significance. The order of significance for the main effects related to Fe grade was determined to be B > A.

Table 7 shows that the F-value of 23.64 signifies the significance of the Fe recovery model, with only a 0.07% probability that such a large F-value could arise from random noise. Additionally, the "Adeq Precision" ANOVA value of 11.6 exceeds 4, demonstrating that the model is sufficient for navigating the design space.

The analysis indicated that the terms A and A^2 were significant in the Fe recovery model (*p*-value < 0.05). According to the F-value, the order of significance for the main effects and interactions influencing Fe recovery is as follows: $A > A^2 > A \times B > B^2 > B$.

The comparison of actual and predicted Fe grade and recovery values is derived from the regression model Equations (4) and (5), which are presented in Table 8.

Figure 7a,b also illustrates the relationship between the actual and predicted Fe grade and recovery, respectively.

According to Figure 7a, a moderately good fit was observed between the actual and predicted Fe, with a correlation R^2 value of 0.56. However, additional investigations may be necessary to refine the regression model. In contrast, Figure 7b illustrates a strong correlation between the actual and predicted values of Fe recovery, as evidenced by an R^2 value of 0.95. This indicates that the developed iron recovery model is effective and reliable.

The steep slope observed in Figure 8 indicates a more substantial influence of the main effect on the Fe grade, as previously predicted with the ANOVA analysis. Furthermore, based on a first-order linear model developed for Fe grade, there is no evidence of a two-way interaction between the input factors affecting Fe grade.

Table 8. Actual and predicted values of Fe grade and Fe recovery in the magnetization roasting optimization.

Run	Fe Grade (%)		Fe Recovery (%)	
	Actual	Predicted	Actual	Predicted
1	61.91	62.86	11.47	11.16
2	64.45	65.90	16.09	15.85
3	65.03	64.88	8.87	8.93
4	66.23	64.99	16.34	16.11
5	66.75	66.60	16.27	16.50
6	65.92	64.69	16.27	16.13
7	62.11	63.72	15.90	14.43
8	65.84	66.71	16.07	16.27
9	64.13	63.78	15.77	17.11
10	66.23	64.99	16.34	16.11
11	64.71	64.07	8.23	9.46
12	65.55	65.68	9.32	8.89

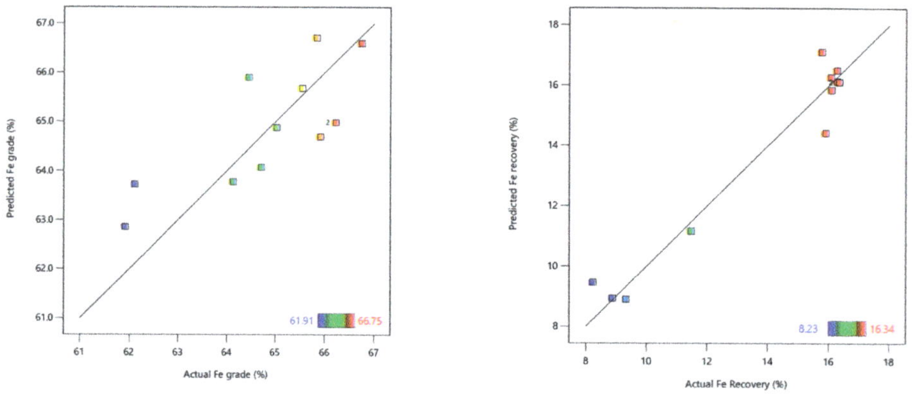

(a) (b)

Figure 7. Magnetization roasting optimization linear correlation plot between the actual versus the predicted (**a**) Fe grade and (**b**) Fe recovery.

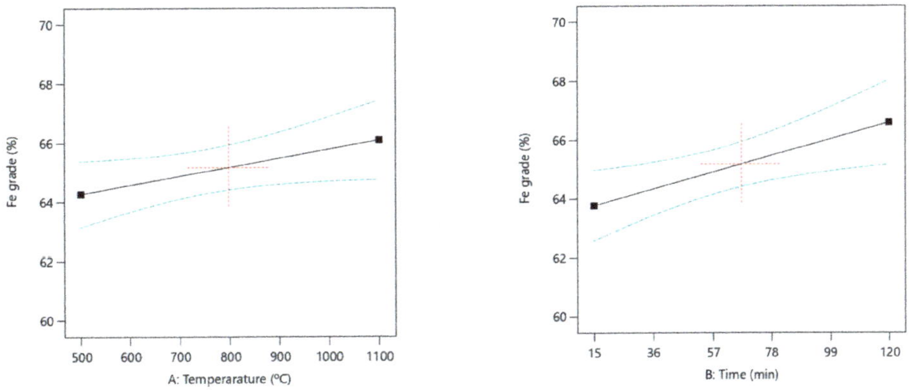

(**a**) (**b**)

Figure 8. Plots showing the main effect of (**a**) roasting temperature and (**b**) roasting time on Fe grade at a confidence level band of 5%.

It was observed that the roasting time is nearly parallel to the horizontal axis, as illustrated in Figure 9b. This indicates that the effect of roasting time on Fe recovery is minimal and consistent with the predictions made with the ANOVA analysis. The interaction plot between roasting temperature and time, also shown in Figure 9c, evaluates the effects of these independent factors. Increasing the roasting temperature improves Fe recovery, irrespective of whether the roasting time is high or low. However, the interaction between roasting temperature and time has a low impact on Fe recovery, as indicated by the ANOVA results. The quadratic effects of temperature and time were not significant, reflected in their low F-values.

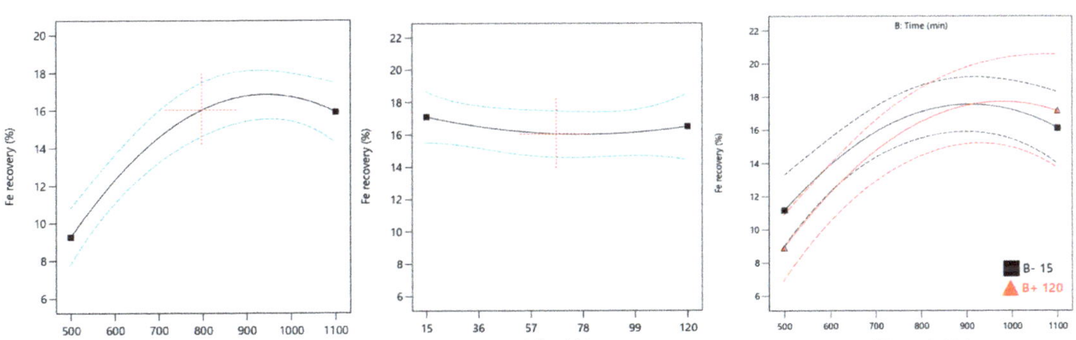

(**a**) (**b**) (**c**)

Figure 9. Plots showing the main effect of (**a**) roasting temperature, (**b**) roasting time, and interaction plots for (**c**) roasting temperature and roasting time on Fe recovery at a confidence level band of 5%.

Three-dimensional response surface and contour plots are essential for comprehending the main and interaction effects of the independent factors. These plots can support decision-making in selecting suitable input conditions to yield the optimal combination for the target response [14,33,34]. The combined contour and 3D surface plots, which display the interactions of the variables on Fe grade and Fe recovery, are presented in Figure 10a,b.

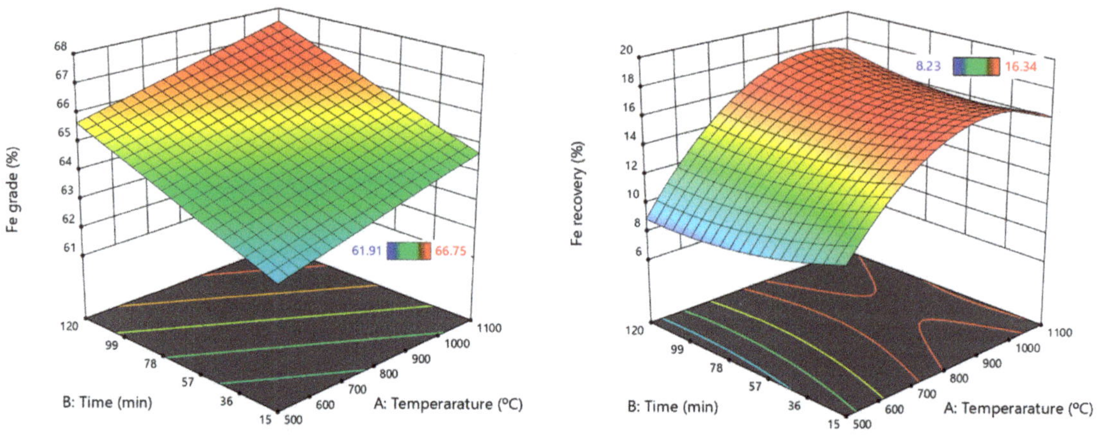

(**a**) (**b**)

Figure 10. Three-dimensional surface plots showing the effect of roasting temperature and time on (**a**) Fe grade and (**b**) Fe recovery.

Utilizing the findings illustrated in Figure 10a,b, the optimum working zone (optimized formulation region) can be easily identified [35]. Figure 10a illustrates that the Fe grade increases with the roasting temperature and time. In contrast, the 3D surface plot regarding Fe recovery indicates that only increased roasting temperature has a positive effect, while additional roasting time does not contribute to improved recovery. Therefore, a shorter roasting time that does not compromise the Fe grade can be chosen to save production time and costs.

In mineral processing, grade and recovery are essential parameters for evaluating the performance of any unit process. As such, the model responses were optimized using the Design-Expert® prediction profiler. During the magnetization roasting process, the optimization aimed to achieve a maximum Fe grade of at least 65%, which is crucial for economic viability. Table 9 summarizes the optimal solution values for variables calculated by the model to achieve the desired Fe grade and Fe recovery. In this context, the desirability function uses a value of "0" to represent an undesirable response and a value of "1" to indicate the most ideal and desirable response.

Table 9. Response prediction for the Fe grade and Fe recovery in the magnetization roasting.

Desirability	Variable		Response	
	A: Temperature (°C)	B: Time (min)	Fe Grade (%)	Fe Recovery (%)
1.000	1050	98	66.76	16.80

The results in Table 9 indicate that operating at optimal roasting temperatures of 1050 °C and a duration of 67 min yields an Fe grade of 66.8%, but only results in a recovery rate of 16.9%. Notably, these roasting conditions lead to the lowest levels of Fe recovery. Therefore, optimizing the next stage, which involves the induced dry-roll magnetic separator, is crucial.

The optimized operating conditions were used to assess the suitability of the developed model in predicting response values. To validate these optimal conditions, an experiment was conducted, resulting in an Fe recovery rate of 17.63% at an Fe grade of

67.45%. The experimental response values aligned well with the predicted values, as illustrated in Table 9. The model is reliable and accurate in forecasting outcomes.

3.4. Optimization of the Induced Dry-Roll Magnetic Separator Process Parameters

The sample generated at the optimum roasting temperature of 1050 °C and a roasting time of 98 min was used to prepare a bulk material for optimizing the induced dry-roll magnetic separation. RSM was also utilized to evaluate the influence of magnetic field intensity (A), rotor speed (B), and product splitter position (C) on iron (Fe) grade and Fe recovery. After conducting the experiments, the outcomes of the experimental design and the observed responses for optimizing induced dry-roll magnetic separation are presented in Table 10.

Two statistical models were developed for each response: Fe grade and Fe recovery. A quadratic second-order polynomial was used to describe the relationship between the input factors and the output responses. The empirical coded equations for both Fe grade and Fe recovery are provided in Equations (6) and (7), respectively.

$$\text{Fe grade (\%)} = 66.02 - 0.15\,A - 0.07\,B + 0.37\,C + 0.72\,A \times B + 0.01\,A \times C + 0.24\,B \times C - 0.02\,A^2 - 0.06\,B^2 - 0.72\,C^2 \quad (6)$$

$$\text{Fe recovery (\%)} = 46.33 - 9.41\,A + 19.77\,B - 4.74\,C - 1.96\,A \times B + 1.52\,A \times C - 1.49\,B \times C - 7.28\,A^2 - 1.59\,B^2 - 6.45\,C^2 \quad (7)$$

where the Fe grade and recovery are given in percentage, magnetic field intensity (A) is in T, rotor speed (B) is in Hz, and product splitter position (C) is in mm.

Table 10. Experimental plan of optimal design with observed values for induced dry-roll magnetic separator optimization.

Run	Variables			Responses	
	A: Magnetic Field Intensity (T)	*B*: Rotor Speed (Hz)	*C*: Product Splitter Position (mm)	Fe Grade (%)	Fe Recovery (%)
1	0.14	70	5	64.99	41.22
2	0.1	70	15	65.53	47.88
3	0.1	40	10	66.37	37.73
4	0.14	60	15	66.35	31.87
5	0.14	20	10	65.06	11.54
6	0.1	70	10	65.84	67.93
7	0.1	70	10	65.82	67.97
8	0.1	40	10	66.38	37.72
9	0.05	40	10	65.31	48.86
10	0.05	70	10	65.61	64.49
11	0.14	60	10	66.1	37.37
12	0.05	20	15	66.53	15.8
13	0.1	20	15	64.98	12.73
14	0.14	20	10	65.09	11.55
15	0.05	70	15	65.17	54.4
16	0.1	40	5	64.81	41.78
17	0.05	20	5	66.3	20.82
18	0.1	40	5	64.84	41.79
19	0.1	40	10	66.37	37.73
20	0.05	70	5	64.12	69.98

The ANOVA analysis was utilized at a 5% significance level to evaluate the adequacy of the developed polynomial regression models for Fe grade and Fe recovery, in which the significance was determined for p-values ≤ 1. Tables 11 and 12 present the ANOVA analysis for Fe grade and recovery, respectively. As shown in Table 11, the F-value of 3.8 indicates that the Fe grade model is statistically significant, with only a 2.46% probability that such an F-value could arise from random noise. Additionally, the "Adeq Precision" ANOVA value of 7.03 exceeds the threshold of 4, suggesting that the model is sufficient and can be effectively used to explore the design space.

Table 11. ANOVA analysis for Fe grade in the induced dry-roll magnetic separation.

Source	Sum of Squares	DF	Mean Square	F-Value	p-Value	Remark
Model	6.99	9	0.7764	3.80	0.0246	Significant
A—Field intensity	0.2285	1	0.2285	1.12	0.3154	not significant
B—Rotor Speed	0.0651	1	0.0651	0.3183	0.5851	not significant
C—Splitter position	1.18	1	1.18	5.77	0.0372	significant
$A \times B$	3.83	1	3.83	18.71	0.0015	significant
$A \times C$	0.0531	1	0.0531	0.2594	0.6216	not significant
$B \times C$	0.3751	1	0.3751	1.83	0.2055	not significant
A^2	0.0014	1	0.0014	0.0070	0.9348	not significant
B^2	0.0103	1	0.0103	0.0502	0.8273	not significant
C^2	2.10	1	2.10	10.27	0.0094	significant
Residual	2.05	10	0.2045			
Cor Total	9.03	19				
Adeq Precision: 7.03						

Table 12. ANOVA analysis for Fe recovery in the induced dry-roll magnetic separation.

Source	Sum of Squares	DF	Mean Square	F-Value	p-Value	Remark
Model	6638.10	9	737.57	40.16	<0.0001	significant
A—Field intensity	901.54	1	901.54	49.08	<0.0001	significant
B—Rotor Speed	4660.62	1	4660.62	253.74	<0.0001	significant
C—Splitter position	194.15	1	194.15	10.57	0.0087	significant
$A \times B$	28.35	1	28.35	1.54	0.2424	not significant
$A \times C$	12.37	1	12.37	0.6733	0.4310	not significant
$B \times C$	13.92	1	13.92	0.7581	0.4044	not significant
A^2	215.41	1	215.41	11.73	0.0065	significant
B^2	6.72	1	6.72	0.3657	0.5588	not significant
C^2	167.49	1	167.49	9.12	0.0129	significant
Residual	183.68	10	18.37			
Cor Total	6821.77	19				
Adeq Precision: 19.69						

The significant factors affecting Fe grade were C, $A \times B$, and C^2 (p-value < 0.05). Based on the F-values, the order of significance for the main and interaction effects on Fe grade is as follows: $A \times B > C^2 > C > B \times C > A > B > A \times C > B^2 > A^2$.

The results from Table 12 indicate that the F-value of 40.16 demonstrates the significance of the developed model for Fe recovery. There is only a 0.01% chance that an F-value this high could occur due to random noise. Furthermore, the "Adeq Precision" ANOVA value of 19.69 surpasses the threshold of 4, confirming that the model is adequate and can effectively guide the design space.

The analysis revealed that the terms A, B, C, A^2, and C^2 are significant (p-value < 0.05). The order of significance of the main and interaction effects was $B > A > A^2 > C > C^2 > A \times B > A \times C > B \times C > B^2$.

The actual and predicted Fe grade and Fe recovery values were compared using the quadratic polynomial Equations (6) and (7), which are given in Table 13. Figure 11a,b illustrates the relationships between the actual and predicted Fe grade and Fe recovery, respectively.

Table 13. Actual and predicted values of Fe grade and Fe recovery in the induced dry-roll magnetic separator optimization.

Run	Fe Grade (%)		Fe Recovery (%)	
	Actual	Predicted	Actual	Predicted
1	64.99	65.00	41.22	44.12
2	65.53	65.85	47.88	50.65
3	66.37	65.99	37.73	41.22
4	66.35	66.11	31.87	29.19
5	65.06	65.14	11.54	10.24
6	65.84	65.94	67.93	63.16
7	65.82	65.94	67.97	63.16
8	66.38	65.99	37.72	41.22
9	65.31	66.30	48.86	44.05
10	65.61	65.29	64.49	68.61
11	66.10	66.21	37.37	39.75
12	66.53	66.18	15.80	13.91
13	64.98	65.34	12.73	14.52
14	65.09	65.14	11.55	10.24
15	65.17	65.08	54.40	54.41
16	64.81	64.94	41.78	39.05
17	66.30	66.13	20.82	23.47
18	64.84	64.94	41.79	39.05
19	66.37	65.99	37.73	41.22
20	64.12	64.05	69.98	69.91

Figure 11a indicates a good fit between the actual and predicted Fe grade values, as evidenced by an R^2 value of 0.77. Meanwhile, Figure 11b suggests a strong correlation between the predicted and experimental data for Fe recovery. This indicates that the Fe grade and recovery models are effective and reliable.

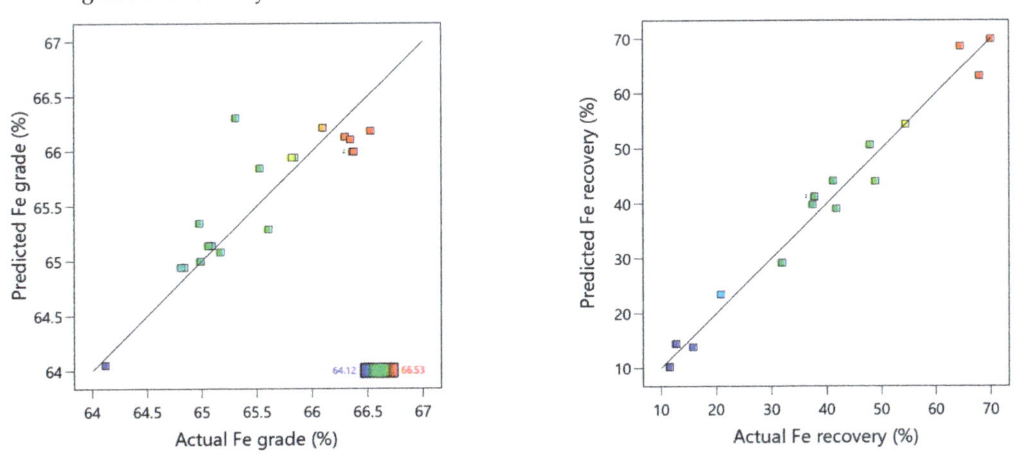

(a) (b)

Figure 11. Induced dry-roll magnetic separator optimization linear correlation plot between the actual versus the predicted (**a**) Fe grade and (**b**) Fe recovery.

However, it is essential to note that one data point (run 8) in the Fe grade analysis lies outside the acceptable range, indicating a residual more significant than 4. This suggests that run 8 is an outlier, and repeating this run could enhance the model's accuracy.

The main and interaction effect plots were developed to illustrate the relationship between the input parameters (magnetic field intensity, rotor speed, and product splitter position) and the output parameters (Fe grade and Fe recovery). The generated plots are presented in Figures 12 and 13 for Fe grade and Fe recovery, respectively.

The results indicate that the product splitter position positively impacts the obtained Fe grade, as evidenced by the steep slopes observed in the main plot for the splitter position. In contrast, the main plots for magnetic field intensity and rotor speed are nearly parallel to the horizontal axis, suggesting that increases in these parameters have an insignificant effect on the Fe grade. This aligns with the ANOVA analysis, which found the *p*-values for field intensity and rotor speed to be insignificant.

The interaction effects involving field intensity and rotor speed, field intensity and splitter position, and rotor speed and splitter position on Fe grade are illustrated in Figure 12. It was observed that the Fe grade increases only with higher field intensity and rotor speed. Lower rotor speeds resulted in poor Fe grades. The interaction between field intensity and splitter position contributed minimally to the Fe grade, as indicated by a plot that is parallel to the horizontal axis. However, operating the equipment at a higher splitter position leads to better Fe grades. Additionally, increasing rotor speed while using a lower splitter position negatively affects the Fe grade. The interactions between field intensity and splitter position and those between rotor speed and splitter position demonstrate minimal impact on the Fe grade due to low F-values, as shown in the ANOVA analysis.

Based on Figure 13, field intensity and rotor speed positively influence Fe recovery, as indicated by the steep slopes in the main effect plots. The Fe recovery increases with higher rotor speeds, while an increase in field intensity has a negative impact on recovery, suggesting that optimal results are achieved at lower field intensities. The main plot for the splitter position remains nearly parallel to the horizontal axis, indicating that adjustments to the splitter position have minimal effect on Fe recovery. This observation is consistent with the ANOVA analysis, which reveals a low F-value for the splitter position.

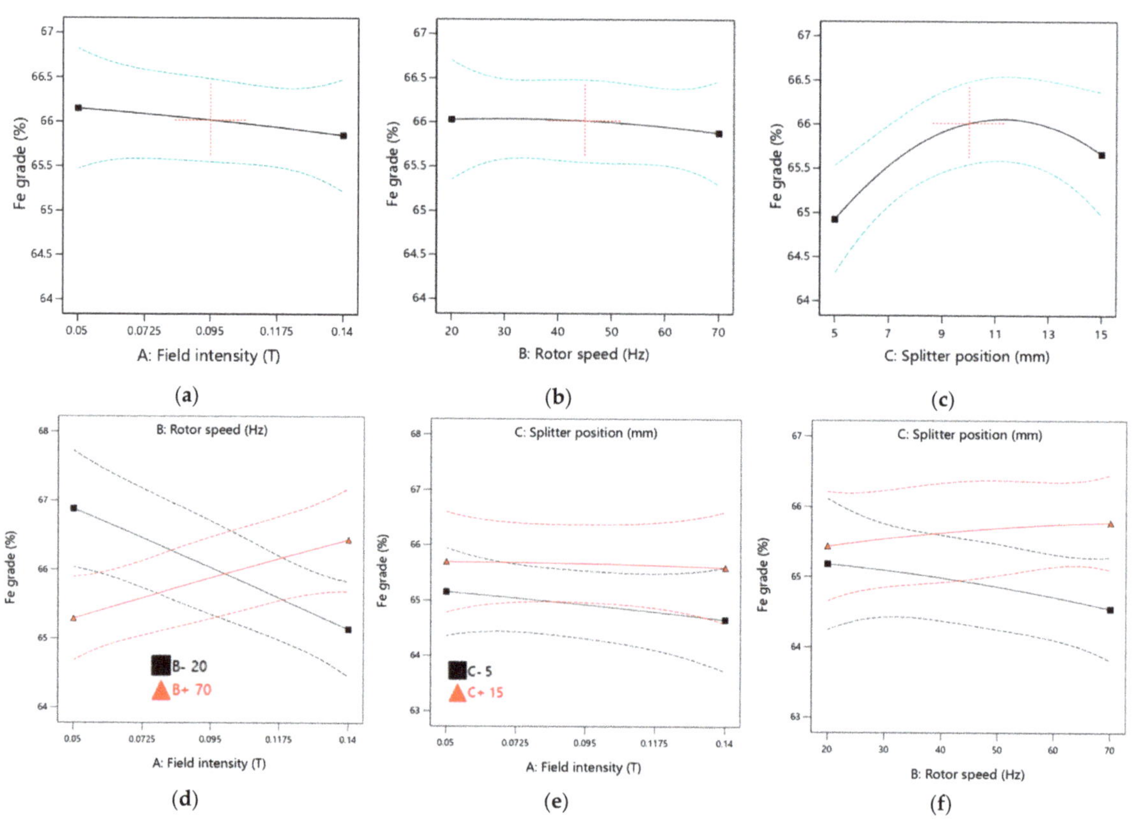

Figure 12. Plots showing the main effect of (**a**) field intensity, (**b**) rotor speed, (**c**) splitter position, and interaction plots for (**d**) field intensity and rotor speed, (**e**) field intensity and splitter position, and (**f**) rotor speed and splitter position on Fe grade at a confidence level band of 5%.

Interaction effect plots for field intensity and rotor speed, field intensity and splitter position, and rotor speed and splitter position on Fe recovery are shown in Figure 13. It is noted that increasing field intensity adversely affects Fe recovery. However, operating the equipment at lower field intensity and rotor speed produces better Fe recoveries. Operating the equipment at a lower rotor speed is not advisable, resulting in significantly lower Fe recovery. The interaction between field intensity and splitter position has a moderate effect on Fe recovery, with optimal conditions identified as low field intensity and low splitter position. Increasing rotor speed while maintaining a lower splitter position enhances Fe recovery values. Nonetheless, all interaction effects have minimal influence on Fe recovery, as indicated by low F-values, with all corresponding p-values exceeding 0.05, which aligns with the ANOVA analysis data.

The 3D surface plots illustrating the interactions of variables affecting Fe grade and Fe recovery are presented in Figures 14 and 15.

According to Figure 14a, the range that promotes a higher Fe grade is found with a splitter position between 10 mm and 11.5 mm at any rotor speed. However, selecting a lower rotor speed is advisable for economic reasons, as it can reduce energy costs. This is further supported by Figure 14b, which indicates that operating the equipment at a low rotor speed and in a lower field intensity region yields optimal Fe grade values.

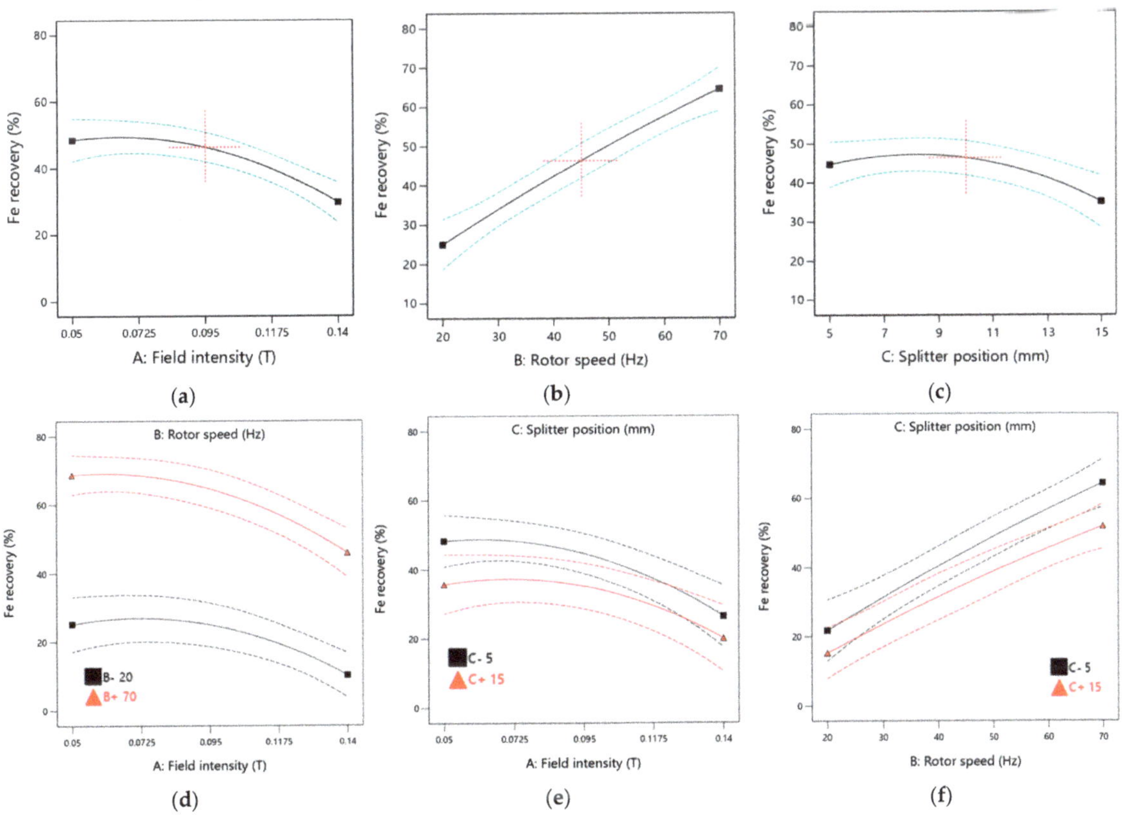

Figure 13. Plots showing the main effect of (**a**) field intensity, (**b**) rotor speed, (**c**) splitter position and interaction plots for (**d**) field intensity and rotor speed, (**e**) field intensity and splitter position, and (**f**) rotor speed and splitter position on Fe recovery at a confidence level band of 5%.

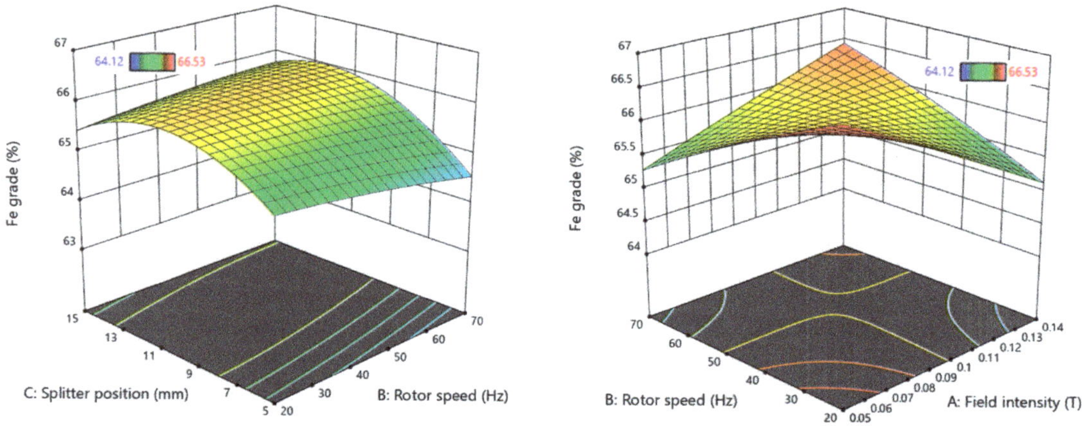

(**a**) (**b**)

Figure 14. Three-dimensional surface plots showing the effect of (**a**) splitter position and rotor speed and (**b**) rotor speed and field intensity on Fe grade.

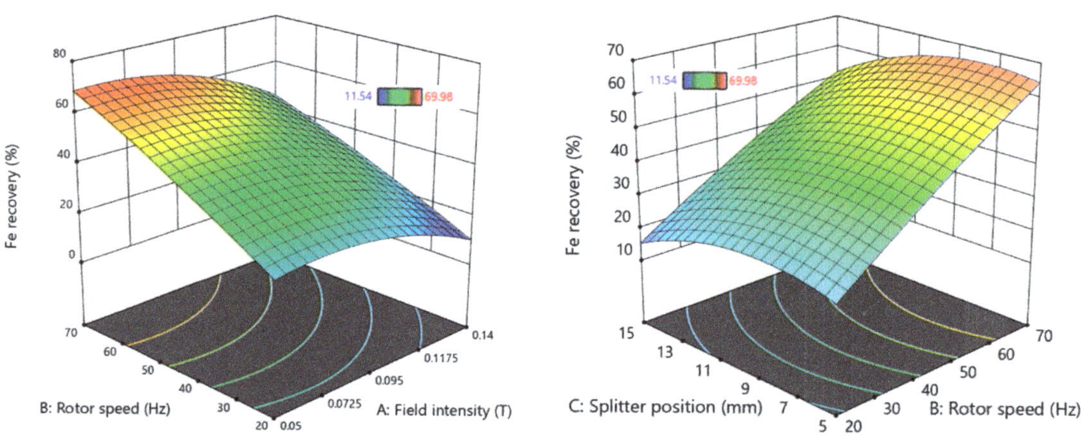

(**a**) (**b**)

Figure 15. Three-dimensional surface plots showing the effect of (**a**) field intensity and rotor speed and (**b**) splitter position and rotor speed on Fe recovery.

In Figure 15a, the conditions for improved Fe recovery are highlighted, showing that a high rotor speed above 65 Hz is necessary while maintaining a field intensity of less than 0.09 T. Figure 15b reveals that, for higher Fe recovery values, the product splitter position should be set below 11 mm at rotor speeds exceeding 65 Hz.

Table 14 presents the best solutions for the variables necessary to achieve the desired Fe grade and recovery. Based on Table 14, the optimum working conditions include a magnetic field intensity of 0.105 T, a rotor speed of 70 Hz, and a product splitter position of 11 mm, resulting in an Fe grade of 66.2% and a recovery rate of 60.2%.

Table 14. Response prediction for the Fe grade and Fe recovery in the induced dry-roll magnetic separation.

Desirability	Variable			Response	
	A: Field Intensity (T)	B: Rotor Speed (Hz)	C: Splitter Position (mm)	Fe Grade (%)	Fe Recovery (%)
0.828	0.105	70	11	66.10	60.20

To validate these optimal conditions, an experiment was conducted, resulting in an Fe recovery rate of 58.29% and an Fe grade of 65.33%. The experimental response values aligned well with the predicted values, as shown in Table 14.

4. Conclusions

The development and optimization of procedures are essential steps in analytical methods. This study aimed to establish optimized working conditions for magnetization roasting and induced dry-roll magnetic separation to achieve the desired Fe grade and recovery. The RSM was employed to model and optimize various operational parameters based on a selected optimal design configuration. The statistical model predicted a satisfactory Fe grade of 66.76%, an increase from an initial 57.1%, with an Fe recovery of 16.7%. This was achieved at a roasting temperature of 1050 °C over 97 min during the magnetization roasting optimization. However, the Fe recovery was notably low, highlighting the need to optimize the induced dry-roll magnetic separator. For the induced dry-roll magnetic separator, the statistical model predicted an optimum Fe grade of 66.1% with an improved Fe recovery of 60.2%. The optimal working conditions were identified as a field intensity of 0.105 T, a rotor speed of 70 Hz, and a product splitter position of 11 mm. Based on the ANOVA analysis, it was found that the optimization of Fe grade during magnetization roasting was significantly influenced by roasting time.

In contrast, the impact of roasting temperature on Fe grade was less pronounced. However, the roasting temperature was crucial in achieving high Fe recovery during magnetization roasting. In the case of the induced dry-roll magnetic separator, only the product splitter position significantly impacted the Fe grade. In contrast, all parameters significantly affected the Fe recovery, with rotor speed having the most significant impact, followed by field intensity. Applying the RMS optimal design has led to developing a robust and reliable method for accurately determining the optimal operating conditions in magnetization roasting and induced dry-roll magnetic separation.

This study presents positive outcomes on the use of biomass as a reductant; however, challenges may arise when scaling this solution. Unlike coal or coke, biomass may struggle with issues of limited supply. Thus, it is essential to conduct further investigations into various agro-industrial biomass sources, such as corn straw, rice husks, sawdust, starch, macadamia nutshells, coconut shells, and cow dung, whether used individually or in combination. Furthermore, the model developed in this study can act as an initial framework for exploring the potential of biomass in these applications.

Author Contributions: Conceptualization, A.S.; methodology, A.S.; writing—original draft preparation, A.S.; investigation, A.S.; data curation, A.S.; supervision, W.N.; writing—review and editing, W.N. All authors have read and agreed to the published version of the manuscript.

Funding: This research received no external funding.

Data Availability Statement: Data can be provided upon request.

Acknowledgments: The authors express their gratitude to the Extraction Metallurgy Department at the University of Johannesburg for granting access to their laboratories and providing invaluable assistance throughout the characterizations and experiments presented in this study. Additionally, the authors would like to thank Assmang Iron Ore for generously supplying the iron plant tailings utilized in this research.

Conflicts of Interest: The authors declare that they have no known competing financial interests or personal relationships that could have influenced the work reported in this paper.

Abbreviations

The following abbreviations are used in this manuscript:

RSM	Response Surface Methodology
Fe	Iron (chemical symbol)
ANOVA	Analysis of Variance
IOT	Iron Ore Tailings
Na_2CO_3	Sodium Carbonate
CO	Carbon Monoxide
CO_2	Carbon Dioxide
CH_4	Methane
H_2	Hydrogen
Al_2O_3	Aluminum Oxide
SiO_2	Silicon Dioxide
O_2	Oxygen
Fe_3O_4	Magnetite
γ-Fe_2O_3	Maghemite
CV	Calorific Value
VM	Volatile Matter
FC	Fixed Carbon
CHNS	Carbon, Hydrogen, Nitrogen, Sulfur (elemental analysis)
XRF	X-ray Fluorescence
XRD	X-ray Diffraction
SEM	Scanning Electron Microscopy
EDS	Energy Dispersive X-ray Spectroscopy
DoE	Design of Experiment
Hz	Hertz (unit of frequency)
T	Tesla (unit of magnetic field intensity)
KVA	Kilovolt-Amperes
Hz	Hertz
mm	Millimeter
MJ/kg	Megajoules per Kilogram
DF	Degrees of Freedom
R^2	Coefficient of Determination

References

1. Zhao, J.; Ni, K.; Su, Y.; Shi, Y. An evaluation of iron ore tailings characteristics and iron ore tailings concrete properties. *Constr. Build. Mater.* **2021**, *286*, 122968. [CrossRef]
2. Liu, S.; Zhao, Y.; Wang, W.; Wen, S. Beneficiation of a Low-Grade, Hematite-Magnetite Ore in China. *Min. Metall. Explor.* **2014**, *31*, 136–142. [CrossRef]
3. Patra, S.; Pattanaik, A.; Rayasam, V. Characterisation of low-grade Barsua iron ore fines and identification of possible beneficiation strategies. *Can. Metall. Q.* **2019**, *58*, 28–45. [CrossRef]
4. Dauce, P.D.; Castro, G.B.D.; Lima, M.M.F.; Lima, R.M.F. Characterisation and magnetic concentration of an iron ore tailings. *J. Mater. Res. Technol.* **2019**, *8*, 1052–1059. [CrossRef]
5. Yang, C.; Li, S.; Zhang, C.; Bai, J.; Guo, Z. Application of Superconducting High Gradient Magnetic Separation Technology on Silica Extraction From Iron Ore Beneficiation Tailings. *Miner. Process. Extr. Metall. Rev.* **2018**, *39*, 44–49. [CrossRef]
6. Peng, T.; Gao, X.; Li, Q.; Xu, L.; Luo, L.; Xu, L. Phase transformation during roasting process and magnetic beneficiation of oolitic-iron ores. *Vacuum* **2017**, *146*, 63–73. [CrossRef]

7. Zhao, B.; Gao, P.; Tang, Z.; Zhang, W. minerals The Efficient Improvement of Original Magnetite in Iron Ore Reduction Reaction in Magnetization Roasting Process and Mechanism Analysis by In Situ and Continuous Image Capture. *Minerals* **2021**, *11*, 645. [CrossRef]
8. Chandio, A.D.; Channa, I.A.; Shaikh, A.A.; Madad, S.; Rizvi, S.B.H.; Shah, A.A.; Ashfaq, J.; Ali Shar, M.; Alhazaa, A. Beneficiation of Low-Grade Dilband Iron Ore by Reduction Roasting. *Metals* **2023**, *13*, 296. [CrossRef]
9. Yu, J.; Han, Y.; Li, Y.; Gao, P. Beneficiation of an iron ore fines by magnetization roasting and magnetic separation. *Int. J. Miner. Process.* **2017**, *168*, 102–108. [CrossRef]
10. Yu, J.; Han, Y.; Li, Y.; Gao, P.; Sun, Y. Separation and recovery of iron from a low-grade carbonate-bearing iron ore using magnetizing roasting followed by magnetic separation. *Sep. Sci. Technol.* **2017**, *52*, 1768–1774. [CrossRef]
11. Suthers, S.P.; Nunna, V.; Tripathi, A.; Douglas, J.; Hapugoda, S. Experimental study on the beneficiation of low-grade iron ore fines using hydrocyclone desliming, reduction roasting and magnetic separation. *Miner. Process. Extr. Metall.* **2014**, *123*, 212–227. [CrossRef]
12. Srivastava, R.; Prasad, R. Mineralogical Characterization-Assisted Reduction Roasting of Banded Hematite Jasper Ore. *J. Inst. Eng. (India) Ser. D* **2020**, *101*, 247–252. [CrossRef]
13. Monzavi, M.; Raygan, S. Beneflciation of an Oolitic-Iron ORE by Magnetization Roasting and Magnetic Separation. *Iran. J. Mater. Sci. Eng.* **2020**, *17*, 17–29. [CrossRef]
14. Rath, S.S.; Sahoo, H.; Dhawan, N.; Rao, D.S.; Das, B.; Mishra, B.K. Optimal Recovery of Iron Values from a Low Grade Iron Ore using Reduction Roasting and Magnetic Separation. *Sep. Sci. Technol.* **2014**, *49*, 1927–1936. [CrossRef]
15. Roy, S.K.; Nayak, D.; Rath, S.S. A review on the enrichment of iron values of low-grade Iron ore resources using reduction roasting-magnetic separation. *Powder Technol.* **2020**, *367*, 796–808. [CrossRef]
16. Sahu, S.N.; Meikap, B.C.; Biswal, S.K. Magnetization roasting of waste iron ore beneficiation plant tailings using sawdust biomass; A novel approach to produce metallurgical grade pellets. *J. Clean. Prod.* **2022**, *343*, 130894. [CrossRef]
17. Chen, H.; Zheng, Z.; Shi, W. Investigation on the kinetics of iron ore fines reduction by CO in a micro-fluidized bed. *Procedia Eng.* **2015**, *102*, 1726–1735. [CrossRef]
18. Cao, Y.; Sun, Y.; Gao, P.; Han, Y.; Li, Y. Mechanism for suspension magnetization roasting of iron ore using straw-type biomass reductant. *Int. J. Min. Sci. Technol.* **2021**, *31*, 1075–1083. [CrossRef]
19. Swickrath, M.J.; Anderson, M. The Development of Models for Carbon Dioxide Reduction Technologies for Spacecraft Air Revitalization. In Proceedings of the 42nd International Conference on Environmental Systems, San Diego, CA, USA, 15–19 July 2012.
20. Roy, S.K.; Nayak, D.; Dash, N.; Dhawan, N.; Rath, S.S. Microwave-assisted reduction roasting—Magnetic separation studies of two mineralogically different low-grade iron ores. *Int. J. Miner. Metall. Mater.* **2020**, *27*, 1449–1461. [CrossRef]
21. Mathieson, J.G.; Somerville, M.A.; Deev, A.; Jahanshahi, S. *Utilization of Biomass as an Alternative Fuel in Ironmaking*; Elsevier: Cambridge, UK, 2015; pp. 581–613. [CrossRef]
22. Ye, L.; Peng, Z.; Wang, L.; Anzulevich, A.; Bychkov, I.; Tang, H.; Rao, M.; Zhang, Y.; Li, G.; Jiang, T. Preparation of core-shell iron ore-biochar composite pellets for microwave reduction. *Powder Technol.* **2018**, *338*, 365–375. [CrossRef]
23. Uy, M.; Telford, J.K. Optimization by Design of Experiment Techniques. In Proceedings of the 2009 IEEE Aerospace Conference, Big Sky, MT, USA, 7–14 March 2009.
24. Lamidi, S.; Olalere, R.; Yekinni, A.; Adesina, K. Design of Experiments (DOE): Applications and Benefits in Quality Control and Assurance. *Intech* **2016**, *11*, 13.
25. Rajewski, J.; Dobrzyńska-Inger, A. Application of response surface methodology (Rsm) for the optimization of chromium(iii) synergistic extraction by supported liquid membrane. *Membranes* **2021**, *11*, 854. [CrossRef] [PubMed]
26. Kincl, M.; Turk, S.; Vrečer, F. Application of experimental design methodology in development and optimization of drug release method. *Int. J. Pharm.* **2005**, *291*, 39–49. [CrossRef]
27. Bowden, G.D.; Pichler, B.J.; Maurer, A. A Design of Experiments (DoE) Approach Accelerates the Optimization of Copper-Mediated 18F-Fluorination Reactions of Arylstannanes. *Sci. Rep.* **2019**, *9*, 11370. [CrossRef]
28. Sabry, I.; Hewidy, A.M.; Alkhedher, M.; Mourad, A.H.I. Analysis of variance and grey relational analysis application methods for the selection and optimization problem in 6061-T6 flange friction stir welding process parameters. *Int. J. Lightweight Mater. Manuf.* **2024**, *7*, 773–792. [CrossRef]
29. Das, B.; Rath, S.S. Existing and New Processes for Beneficiation of Indian Iron ores. *Trans. Indian Inst. Met.* **2020**, *73*, 505–514. [CrossRef]
30. Wills, B.A.; Napier-Munn, T. *Mineral Processing Technology: An Introduction to the Practical Aspects of Ore Treatment and Mineral*, 7th ed.; Elsevier: Oxford, UK, 2006. [CrossRef]
31. Malenga, E.N.; Mulaba-Bafubiandi, A.F.; Nheta, W. Application of the response surface method (RSM) based on central composite design (CCD) and design space (DS) to optimize the flotation and the desliming conditions in the recovery of PGMs from mine sludge. *Sep. Sci. Technol.* **2022**, *57*, 2960–2983. [CrossRef]

32. Ramudzwagi, M.; Tshiongo-Makgwe, N.; Nheta, W. Optimization of Hydrocyclone Classifier in the Beneficiation of Fine Coal Using Taguchi Approach. Ph.D. Thesis, University of Johannesburg, Johannesburg, Zuid-Afrika, 2020.
33. Modiga, A.; Abdulsalam, J.; Eterigho-Ikelegbe, O.; Bada, S.; Goso, X. Optimisation of ultrasound-assisted alkali-acid leaching of rare earth elements from run-of-mine and discard coal using the response surface methodology. *Miner. Eng.* **2023**, *204*, 108377. [CrossRef]
34. Sadi, J.A. Designing experiments: 3 level full factorial design and variation of processing parameters methods for polymer colors. *Adv. Sci. Technol. Eng. Syst.* **2018**, *3*, 109–115. [CrossRef]
35. Nunna, V.; Hapugoda, S.; Eswarappa, S.G.; Raparla, S.K.; Kumar, R.; Nanda, N.K. Beneficiation of Low-grade Iron Ore Fines by Using a Circulating-type Air Classifier. *Miner. Process. Extr. Metall. Rev.* **2019**, *40*, 356–367. [CrossRef]

minerals

MDPI

Article

Financial and Technological Potential of Eco-Efficient Recycling of Waste Electronic Equipment

Tomasz Suponik [1,*], Paweł Friebe [2,*], Umut Kar [3], Dawid M. Franke [1] and Paulina Gołuch [1]

1 Department of Geoengineering and Resource Exploitation, Silesian University of Technology, 44-100 Gliwice, Poland; dawid.franke@polsl.pl (D.M.F.); paulgol658@student.polsl.pl (P.G.)
2 KOMAG Institute of Mining Technology, 44-101 Gliwice, Poland
3 Department of Mining and Metallurgical Engineering, University of Nevada, Reno 1664 N. Virginia St, Reno, NV 89557, USA; ukar@unr.edu
* Correspondence: tomasz.suponik@polsl.pl (T.S.); pfriebe@komag.eu (P.F.)

Abstract: The paper presents the financial potential of recycling waste electronic equipment (WEE) in the form of printed circuit boards, hard drives, and lithium-ion batteries. Metal contents in selected types of WEE were presented, as well as their price and importance from an environmental, economic, and geopolitical perspective using indicators of relative supply risk and abiotic depletion potential (ADP). The potential benefits that recycling can bring to the company and the environment were presented. Furthermore, the mass balance and value of recovered metals were estimated for southern Poland, and potential possibilities for the management of the remaining separation products were presented. Finally, verified physical recycling methods for the presented WEE were described.

Keywords: recycling of waste electronic equipment (WEE); metal recovery; financial and technological aspects

1. Introduction

Waste electronic equipment (WEE) represents a group of wastes whose quantity is growing dynamically from year to year and, due to the presence of metals, which are often critical, and plastics, constitutes a global environmental, economic, and even political problem. Such a rapid rise in end-of-life electronic equipment poses a challenge both for companies seeking to recover valuable raw materials and for authorities striving to reduce the environmental impact of this waste stream.

From a global economic perspective, the dynamic increase in WEE is closely related to the growth of the world's population and the escalating consumption of electronic devices, which together amplify the demand for the metals required to manufacture them. Modern electronic devices are highly dependent on critical raw materials, including rare earth elements (REEs), cobalt, nickel, lithium, and precious metals [1]. The demand for these metals is driven not only by consumer electronics but also by ongoing digital and energy transformations. The digital transition stimulates the development of information and communication technologies (ICTs), widespread use of mobile devices, data centres, and the Internet of Things (IoT) [2]. Meanwhile, the energy transition requires the implementation of low-emission power sources (e.g., wind turbines, photovoltaic cells) and supports the development of electromobility [3]. For example, lithium, cobalt, and REEs are crucial in the manufacturing of batteries for electric vehicles, whereas copper is indispensable for power grids and wind turbine components.

The most commonly used methods for recovering metals from small-sized WEE are chemical methods that involve their dissolution in acids or incineration. In addition to the high complexity of these methods and the high emissions to the environment, there is a loss of non-metallic parts that could be used for other production or manufacturing processes. Furthermore, the process costs and installation costs of these methods are often very high. Alternative methods for processing WEE are mineral engineering methods. They allow for the separation of various materials, including plastics, which can then be reused to create the same or other products without creating excess emissions into the environment. We can then say that this is an eco-efficient WEE recycling technology that is in line with the idea of a circular economy and sustainable production.

It should be noted that products obtained by purely physical methods typically do not achieve elemental purity but rather are metal-rich concentrates. Nevertheless, the authors' research demonstrated exceptions, such as obtaining a high-purity neodymium alloy concentrate directly from HDDs, which was successfully processed into fully functional magnets without chemical treatments, and producing economically valuable metal-rich concentrates from WPCBs suitable for further metallurgical refining.

This article focusses on the financial and technological potential of eco-efficient WEE recycling. Three groups of small-sized WEE were analysed: printed circuit boards (WPCBs), computer hard drives (WHDs), and lithium-ion batteries (WLIBs). The subsequent sections present the following:

- The content and price of metals in selected types of WEE, along with an overview of indicators that evaluate their importance from environmental, economic, and geopolitical perspectives, particularly relative supply risk and abiotic depletion potential (ADP);
- Price trends of metals used to create electronic equipment from 2000 to 2024;
- The mass balance and estimated value of recoverable metals in WEE collected in southern Poland, illustrating the economic and environmental potential of recycling;
- Review of the research on eco-efficient recycling of WEE conducted by the authors of the article against the background of other methods of recovering valuable substances.

WPCBs, WHDs, and WLIBs were selected due to their high content of critical and valuable metals and diverse material composition, and because they represent three technologically distinct recycling challenges for which the authors have developed and experimentally verified dedicated recovery processes.

2. Methods

The literature analysis method was used to determine the prices of metals used in the construction of electronic equipment and their average content in the WEE, as well as to analyse the indicators of relative supply risk and abiotic depletion potential (ADP). Based on the metal prices in March 2025 and the average metal content in the relevant WEE, the estimated metal prices in 1 Mg of waste were then calculated.

To assess the importance of metals present in WEE from a geographical, geological, environmental, economic, and political perspective, two indicators were used: relative supply risk and abiotic depletion potential (ADP). The second parameter is used to assess the impact of resource depletion. It focusses specifically on abiotic resources, which are non-living materials such as minerals and metals. ADP describes the ratio of the annual production of a resource to its ultimate reserves (based on the crustal content). It allows an assessment of the extent to which the use of a particular material contributes to the depletion of its resources. The higher the ADP value for a particular resource, the greater the potential for the resource to deplete [4]. Just as the global warming potential (GWP) of various pollutants is converted to CO_2 equivalents, ADP values are commonly converted into antimony equivalents (Sb eq.). The other indicator is the relative supply risk developed by the Royal Society of Chemistry, which assesses future raw material availability issues. The indicator is based on factors such as crustal abundance, distribution and quantity of reserves, production concentration, substitutability, recycling rates, and political stability [5]. The relative supply risk ranges from 1 (very low risk) to 10 (very high risk).

In order to determine the mass balance of WEE and the estimated value of metals in them for the Silesian Voivodeship and, on this basis, for the whole of Poland, data were obtained from three anonymous companies located in the Silesian Voivodeship dealing with WEE collection.

Finally, the results of research conducted by the authors of the article on the recycling of WEE in accordance with the principles of circular economy and sustainable production were presented.

3. Results

3.1. Content, Price, and Environmental, Economic, and Geopolitical Importance of the Metals Present in the WEE

3.1.1. Waste Printed Circuit Boards (WPCBs)

One of the most common types of WEE is waste printed circuit boards (WPCBs), which are made of a composite material with embedded metal tracks as the base. The most commonly used PCBs are of type FR-4, whose composite is made of glass fibre in an epoxy resin matrix with flame retardant additives. Copper is the most abundant metal in PCBs, but metals such as aluminium, lead, nickel, tin, and zinc are also present, as well as small amounts of precious metals such as gold, silver, platinum, and palladium [6,7] (Table 1). The metal content in PCBs is tens or even hundreds of times higher than that of ores [8]. As shown in Table 1, the WPCBs consist of metals with medium (Cu, Al, Zn, Au), increased (Pb, Ni, Sn, Ag), and high (Pd, Pt) relative supply risk. For example, lead (Pb) has a relative supply risk value of 6.2, according to the Royal Society of Chemistry. Although Pb is widely recycled and geologically abundant, its risk level is elevated due to its dependence on byproduct extraction (mainly from zinc ores), declining primary production, and regulatory constraints.

In terms of abiotic depletion potential (ADP), the values are very high for Au, Pd, and Pt; medium for Ag; and low or very low for Cu, Al, Pb, Zn, Ni, and Sn. This means that resources of some metals such as Au, Pd, and Pt are particularly limited, and that their availability is further threatened by geopolitical or production factors.

Both indicators show that the key metals from an economic, environmental, and political perspective are Pd, Pt, Au, and Ag. This is reflected in their market value. Therefore, the most valuable metals in WPCBs—considering both their content and price—are gold, silver, palladium, and, due to high abundance, copper (Table 1). The recovery of these metals can generate significant economic benefits for recycling companies [9]. The estimated value of metals in 1 Mg (1000 kg) of WPCBs is nearly USD 92,000, with gold accounting for the largest portion. Metals are treated as target concentrates, whereas plastics are treated as residual materials for which downstream applications are sought.

3.1.2. Waste Hard Drives (WHDs)

Another type of WEE is waste hard drives (WHDs), which have an average weight of 509.8 g and consist of several elements [13]. These include aluminium, steel, plastic, printed circuit boards, and neodymium magnets. Most of the mass share belongs to the HDD casing, which consists of two parts: the main part and its cover. Typically, this casing is made of an Al-Si alloy. The HDD also consists of a disc stack assembly, which includes platters with a spacer and a spindle cover. The platters are made of an Al core with a thin layer of metal alloy providing magnetic properties. This alloy contains Co, Ni, Fe, and

Zn [14]. The HDD also contains a PCB that is mounted on the bottom of the case. The HDD also consists of a Voice Coil Assembly (VCA), which is made up of neodymium magnets, ferrous metal holders, and other mechanical parts. The VCA contains metals such as Fe, Nd, Pr, Dy, and Co [15]. An HDD is composed of about 83% (by weight) metals and about 17% non-metallic assemblies such as PCBs and plastic parts. The Al content is the highest because the HDD casing is made of this metal.

Table 1. Estimated metal content of type FR-4 WPCBs and related indicators of relative supply risk and abiotic depletion potential [4,5,7] and estimated value of metals in WPCBs [10–12].

Metal	Average Metal Content in WPCB, %	Relative Supply Risk	ADP, kg Sb eq/kg	Estimated Value of Metal in 1 Mg, USD
Cu	20	4.3	2.1×10^{-2}	1821.6
Al	2	4.8	2.5×10^{-8}	49.5
Pb	2	6.2	1.9×10^{-2}	39.3
Zn	1	4.8	2.8×10^{-3}	32.2
Ni	2	6.2	8.1×10^{-4}	303.6
Sn	4	6.7	8.1×10^{-2}	1194.1
Ag	0.20	6.2	8.6×10	128.4
Au	0.10	5.7	1.4×10^{3}	84,556.5
Pd	0.0050	7.6	9.7×10^{2}	1532.0
Pt	0.0015	7.6	9.7×10^{2}	454.3
Total	31.31	-	-	91,910.9

HDD consists of metals with medium (Al, Fe, Cu, Ti, Zn, Au), increased (Ni, Sn, Ta, Ag), high (Co, Pd, Pd, Pt), and extremely high (Nd, Ce, Pr, Dy, Sb) relative supply risk (Table 2), while the ADP is very high for Au and Pd, medium for Ag and Sb, low for Cu, Ni, Sn, Zn, and Pr, and very low for Ce, Nd, Dy Co, Al, Si, Fe, Ti, and Ta (Table 2), meaning that the resources of metals used in HDD production, apart from those from which PCBs are made, are large, but the availability risk, especially of Nd, Pr, Dy, Ce, and Sb due to production concentration, substitutability, recycling rates, and political stability, is extremely high. The most important metals from an economic, environmental, and political perspective that drive the need for WHD recycling are Nd, Pr, Ce, and Dy, which make up neodymium magnets, and Pd, Au, and Ag, which are found in PCBs that control HDD.

The most valuable metals in WHD, based on their content, are gold, cerium, copper, neodymium, palladium, dysprosium, silver, and praseodymium, which are contained in neodymium magnets and PCBs (Table 2). Aluminium and quartz are also of great value due to their high content in HDD. The HDD casing is made of aluminium with a steel cover. The estimated value of metals contained in a ton of WHD is around USD 9908, with the highest-value metals found in neodymium magnets, PCBs, and HDD casings. Therefore, they are treated as concentrates in the recycling process.

Over the past decade, there has been a marked transition from HDDs to solid-state drives (SSDs) in both consumer and enterprise storage markets. According to a study of the global storage market, total HDD shipments decreased from approximately 650 million units in 2010 to approximately 260 million units in 2020 [17]. Despite the continued surge in SSD adoption, market analyses from 2023 and early 2024 indicate that HDDs remain relevant in specific niches. According to a recent IDC report, worldwide HDD shipments dropped from around 260 million units in 2020 to approximately 180 million units in 2023, while SSD shipments exceeded 400 million units during the same period [17].

However, HDDs continue to be deployed in cost-sensitive environments and in applications requiring high-capacity storage with less stringent input/output performance, such as long-term data archiving, near-line backups, and certain big data analytics platforms where throughput needs can be met by high-density HDD arrays [18]. Additionally, large-scale cloud providers still integrate HDDs for data storage purposes to balance capital expenditures with operational efficiency. Although the price gap between SSDs and HDDs continues to narrow, magnetic disc technology is likely to retain a share of the enterprise storage market for at least the next five years, particularly where massive storage capacity remains a priority. As a result, a dual-market model is expected to persist, with SSDs dominating performance-centric segments and HDDs occupying high-capacity, lower-cost storage roles.

Table 2. Estimated metal content in WHDs and related indicators of relative supply risk and abiotic depletion potential [4,5,14,15] and estimated metal value in WHDs [10–12,16].

Part of WHD	Metal	Average Metal Content in WHD, %	Relative Supply Risk, -	ADP, kg Sb eq/kg	Estimated Value of Metal in 1 Mg, USD
HDD casing	Al	64.0	4.8	2.5×10^{-8}	1583.4
	Si	8.9	-	1.4×10^{-11}	124.3
	Fe	3.2	5.2	5.24×10^{-8}	10.1
PCBs	Cu	5.4	4.3	2.1×10^{-2}	491.6
	Ni	0.18	6.2	8.1×10^{-4}	27.3
	Sn	0.17	6.7	8.1×10^{-2}	50.7
	Ti	0.04	4.8	2.79×10^{-8}	2.6
	Zn	0.03	4.8	2.8×10^{-3}	1.0
	Ce	0.04	9.5	1.5×10^{-5}	1.2
Neodymium magnets	Nd	0.73	9.5	3.1×10^{-5}	438.5
	Pr	0.11	9.5	1.4×10^{-4}	72.4
	Dy	0.08	9.5	7.0×10^{-5}	195.5
	Co	0.07	7.6	1.57×10^{-5}	17.0
PCBs	Sb	0.01	9.0	1.0	1.7
	Ta	0.06	7.1	4.06×10^{-5}	202.2
	Ag	0.0234	6.2	8.6×10	225.6
	Au	0.0071	5.7	1.4×10^{3}	6003.5
	Pd	0.0015	7.6	9.7×10^{2}	459.6
	Total	83.0	-	-	9908.3

3.1.3. Waste Lithium-Ion Batteries (WLIBs)

The last example of WEE presented is waste lithium-ion batteries (WLIBs). LIBs are now the dominant portable power source in smartphones, laptops, cameras, handheld devices, and other high-tech devices. The structure of LIBs is not homogeneous, and multi-metallic compounds can be found in them. Lithium cobalt oxide (LCO) and lithium nickel manganese cobalt oxide (NMC) are examples of the types of active cathode materials (CAM) used in these batteries [19]. They typically contain approximately 21.31% Mn, 16.54% Co, 2.56% Ni, 2.22% Li, 5.93%–14% Cu, 4.7%–9.12% Al, 0.04%–25% Fe, and approximately 15% organic materials and 7% plastics [20,21]. Organics include electrolytes, binders, and separators [19,20]. LIBs consist of metals with medium (Mn, Cu, Al, Fe), elevated (Ni, Li), and high (Co) relative supply risk, while the ADP is low or very low for Cu and Ni and Mn, Co, Li, Al, and Fe (Table 3), respectively, which means that the resources of metals used for battery production are large, but the availability risk, especially of Co, Li, and Ni, due to production concentration, substitutability, recycling rates, and political stability, is quite high. The last indicator makes the value of the metals Li, Ni, and Co significant (Table 3).

Table 3. Estimated metal content in WLIBs and related indicators of relative supply risk and abiotic depletion potential [4,5,20,21] and estimated metal value in WLIBs [10,12].

Metal	Average Metal Content in WPCBs, %	Relative Supply Risk	ADP. kg Sb eq/kg	Estimated Value of Metal in 1 Mg, USD
Mn	21.31	5.7	3.3×10^{-5}	441.1
Co	16.54	7.6	1.57×10^{-5}	4019.2
Ni	2.56	6.2	8.1×10^{-4}	388.6
Li	2.22	6.7	5.2×10^{-5}	383.5
Cu	5.93–14	4.3	2.1×10^{-2}	907.6
Al	4.7–9.12	4.8	2.5×10^{-8}	171.0
Fe	0.04–25	5.2	5.24×10^{-8}	39.7
Total	53.3–90.75	-	-	6350.7

WLIB recycling can also provide significant financial benefits to the recycler. Considering the current metal prices on the exchanges and the average metal content in WLIBs, it is estimated that in one tonne of WLIBs, the value of the metals contained is over USD 6350. Note that cobalt is the key metal determining the profitability of WLIB recycling, although copper, nickel, lithium, and manganese are also high-value metals.

3.2. Metal Price Volatility 2000–2024

Due to digital transformation and the development of information and communication technologies, but also the geopolitical situation, the demand for metals is increasing. Higher demand causes their prices to rise. Therefore, since the beginning of the 21st century, there has been an increase or fluctuation in the prices of many metals on world markets (Figure 1). Price fluctuations are usually due to geopolitical, economic, or technological factors [22]. For example, in the case of cobalt, production in the Congo accounts for 70% of the world market, while the shares of the three largest producers of lithium and REEs were, in 2024, 92% and 90%, respectively. For lithium, these were Australia, Chile, and China, and for REEs, China, the US, and Myanmar [3,23]. Additionally, prices fluctuate according to global economic cycles. In general, since 2000 (for Nd since 2015), there has been an increase in the prices of platinum, palladium, gold, and silver by 75, 33, 756, and 471%, respectively, as well as nickel, aluminium, copper, lithium, and neodymium by 112, 62, 380, 1015, and 31%, respectively. Among the metals analysed, only the price of cobalt dropped by 61%, from around USD 41,000 to USD 20,000, mainly due to geopolitical reasons.

All of these factors, such as the growth of WEE volumes, the increase in metal demand, and volatility, underscore the importance of effective recycling as a means of reducing the pressure on primary resources, mitigating supply risks, and minimising environmental impact [24]. WEE recycling can provide significant financial benefits by stabilising metal supply chains and reducing vulnerability to price fluctuations [25], as well as environmental benefits through lower energy consumption and a reduced environmental footprint compared to traditional mining and smelting [26].

3.3. Mass Balance and Estimated Value of Metals in WEE for Southern Poland

Eco-efficient recycling of WEE can bring significant economic benefits. The mass balance and estimated value of metals in WPCBs, WHDs and WLIBs — both per 1 million inhabitants and extrapolated to the national scale — are summarised in Table 4. In Poland, with a population of 38.4 million, it is estimated that approximately 1045 Mg of WPCBs, 522 Mg of WHDs, and 1742 Mg of WLIBs are collected annually. The estimated value of the metals contained in these waste streams amounts to approximately USD 96 million for WPCBs, USD 5.1 million for WHDs, and USD 11 million for WLIBs.

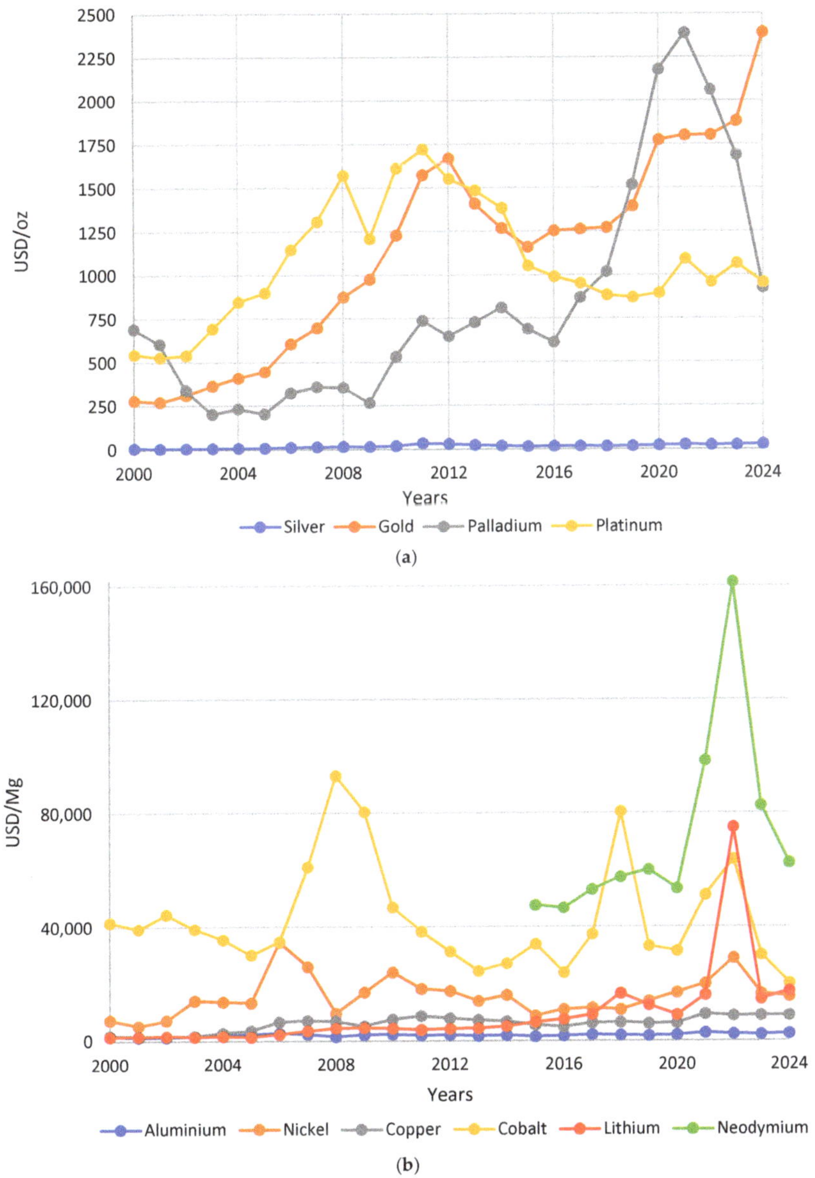

Figure 1. Historical price trends of (**a**) silver, gold, palladium, and platinum [10] and (**b**) nickel, copper, cobalt, lithium, aluminium [10,11], and neodymium [16] between 2000 and 2024.

In addition to regional estimates, it is important to place the value and quantity of waste electronic equipment (WEE) in a global context. According to the Global E-waste Monitor 2024, a record 62 million tonnes of e-waste was generated worldwide in 2022, representing an 82% increase since 2010 and averaging 7.8 kg per person. This figure is projected to rise by another 32%, reaching 82 million tonnes by 2030. However, only 22.3% of the 2022 e-waste mass was formally collected and recycled in an environmentally sound manner, leading to significant losses of secondary resources. The estimated value of metals embedded in the 2022 e-waste stream was approximately USD 91 billion, including USD 19 billion in copper, USD 15 billion in gold, and USD 16 billion in iron [27].

Table 4. Mass balance and estimated value of metals in WEE (WPCBs, WHDs, and WLIBs) for southern Poland in 2024. Data based on averages from three collection companies in the Silesian Voivodeship. Values are provided per 1 million inhabitants (top rows) and extrapolated to the entire Polish population (bottom rows).

Description	Unit	Value for		
		WPCBs	WHDs	WLIBs
Estimated mass of WEE collected (per 1 million inhabitants)	Mg/year	27.2	13.6	45.3
Mass of non-metallic product generated during recycling (per 1 million inhabitants)	Mg/year	19.3	2.3	3.17 (plastics) 6.79 (organic materials)
Estimated value of metals in WEE (per 1 million inhabitants)	USD/year	2,499,669	134,737	287,863
Estimated value of metals in WEE collected in Poland	USD/year	96,080,394	5,178,930	11,064,661

Among the countries generating the largest volumes of e-waste in 2024 were China (10.1 Mt), the United States (6.9 Mt), and India (3.2 Mt). Despite their contribution to the global total, recycling rates in these countries remain low: approximately 16% in China, 15% in the USA, and just 1% in India. By contrast, Germany and the United Kingdom have achieved significantly higher formal collection rates of 52% and 57%, respectively [28]. These disparities underscore the need for harmonised and enforced recycling frameworks as well as the implementation of efficient recovery methods, such as those analysed in this study.

3.4. Research on Recycling of WEE

Currently, WEE recycling is carried out by chemical methods, i.e., hydrometallurgy, pyrometallurgy, and biohydrometallurgy, or by physical methods. Where physical methods can be used in recycling, they should be used in preference to chemical methods, as the use of the latter usually requires special permits and increased work to prevent environmental degradation. It should also be remembered that chemical changes occurring during WEE recycling processes cause the destruction or severe contamination of non-metal parts, which in turn creates additional waste. In the case of physical methods, no hazardous substances or devices that cause emissions into the atmosphere are used, so these methods can be considered more environmentally friendly. Proper recycling using these methods allows for high yields at relatively low costs. Furthermore, plastics obtained as a byproduct of recycling remain chemically unchanged, which significantly increases the potential for further use, e.g., in the production of resin-based composite materials. Note that physical methods often result in the creation of a mixture of metals that must be further chemically processed to isolate the individual metals.

The following section presents an overview of the research conducted by the authors of the paper in the field of WPCB, WHD, and WLIB recycling.

In WPCB recycling, one of the most effective physical methods is electrostatic separation [8,29]. To do this properly, disintegration and grinding are required, preferably using a shredder and a knife mill at cryogenic temperatures. Separation occurs as a result of differences in the surface charge accumulation capacity of the grains and is carried out without the use of process water. Separation results in a mixture of metals and separately ground composite parts. Due to the high purity of the products, the valuable metal mixture can be sold to local metal smelters, while the product consisting of non-metallic parts can be used as a filler in the production of composite materials (Table 5) [8]. Due to this, recycling carried out in this way is relatively simple to implement and does not require the use of complicated technical equipment.

Table 5. Amount of separation products and method of using them for WEE recycling.

Type of WEE	Products Obtained from WEE Recycling	Amount of Separation Products, %	Application or Intended Use of Separation Products	Description of the Directions/Method of Use of the Products
WPCB	Metal mixture (Cu, Au, Ag, Al, Pb, Zn, Ni, Fe, Sn, Pd, Pt)	26.2	Sale of metal mixtures to existing metal production or recovery plants	Financial benefits
	Non-metal products (plastics and glass fibre)	71.0	Product can be used as a filler in the production of composite materials based on polymer materials	e.g., production of resin floors and composite boards, with financial benefits
	Conglomerate of plastics and metals	2.8	Processing by bioleaching using the bacteria *Acidithiobacillus ferrooxidans* [30]	- Long-term kinetic process-method ineffective for large amounts of WPCBs - Environmentally friendly method - Alternatively, this product can be used directly in the chemical process, and due to its small quantity, this should have a relatively smaller negative impact on the environment because there will be less emissions and fewer chemicals will be used
WHD	Fine-grain NdFeB alloy	1.02	Sale of neodymium alloy to produce new NdFeB magnets using the magnet-to-magnet [31] or hydrometallurgical method [32]	Financial benefits, including the production of new NdFeB magnets
	Ferromagnetic steel	6.60	Sales of metals to existing metal production or recovery plants	Financial benefits - Ferromagnetic and stainless steel is sorted, purified, and remelted in furnaces to produce feedstock for new products - Aluminium is usually sorted, decontaminated, and melted, yielding secondary aluminium that requires only about 5% of the energy used in primary production [33]
	Stainless steel	8.94		
	Aluminium	73.01		
	WPCB grains	4.76	Recycling of WPCBs, as described above	Benefits of recycling as stated for WPCBs
	Plastics	3.83	For recycling plastics	Production of new consumer goods
	Mixture of fine metal grains (Cu, Al, Fe, Sn)	1.84	Sale of metal mixtures to existing metal production or recovery plants	Financial benefits
WLIB	Black mass (Li, Co, Ni, Mn, graphite)	40	Hydrometallurgical recovery	Financial benefits
	Aluminium foil	8	Recycled as secondary Al	Valuable Al feedstock
	Copper foil	10	Recycled as secondary Cu	Valuable Cu feedstock
	Plastics and organics (binders, separators)	36	For recycling plastics	
	Steel shells	6	Remelted in steel plants	Feed for steel manufacturing

Recent studies have also demonstrated innovative approaches for the recovery of aluminium and iron from WEE using integrated mechanical and thermal techniques, further confirming the potential for sustainable valorisation of these metals [34,35].

In the investigated WHD recycling technology, both the initial mechanical shredding and the subsequent selective separation processes play a key role [36]. These steps enable the recovery of several fractions of material with different compositions and industrial applications, namely aluminium, steel, neodymium alloy (NdFeB), printed circuit boards (PCBs), and plastics (Table 5). The metal recovery process of hard drives begins with mechanical shredding, which enables the disintegration of components and the release of embedded metals. This is followed by a two-stage magnetic separation process: in the first stage, using a stationary steel plate, NdFeB magnet fragments are separated along with accompanying ferromagnetic contaminants, whereas in the second stage, a strong permanent magnet is employed to separate the remaining ferromagnetic metals, primarily structural steel. The next step involves the granulometric classification of the shredded material, resulting in division into four particle size classes, >25 mm, 25–10 mm, 10–3.15 mm, and 3.15–0 mm, facilitating its selective further processing. The >25 mm particle size class, composed only of stainless steel, is considered a final product and directed to metallurgical recycling. The 25–10 mm fraction undergoes gravity separation on a concentration table, where aluminium is recovered as a product, while the remaining material is subjected to optical separation to extract printed circuit board (PCB) fragments and plastics as distinct products. The 10 to 3.15 mm fraction is subjected to cyclofluidic gravity enrichment in air, resulting in aluminium recovery, after which the remaining material is directed to optical separation for further separation of PCB grains from plastic grains. The finest particle size class, 3.15 to 0 mm, undergoes electrostatic separation, where differences in electrical conductivity allow the recovery of metals and plastics. The method of hard drive recycling has been submitted to the Polish Patent Office [37].

The pretreatment process of WLIBs, which is divided into various steps such as discharge, disintegration/dismantling, comminution, classification, separation, dissolution, and thermal treatment, is of essential importance, particularly in the cases of hydrometallurgical extraction and direct recycling of spent LIBs. The pretreatment process has been shown to improve recovery efficiency and reduce energy consumption in subsequent recycling processes similar to mineral processing for extractive metallurgy [38,39]. The primary objective of the pretreatment is to separate the various components, such as separators, current collectors, etc., from the black mass of the LIBs in a safe and effective manner [40].

The process begins with the discharge of the batteries to eliminate any residual electrical charge, thereby preventing thermal runaways, gas emissions, or fires during further processing. This is followed by dismantling/disintegration, either manually for lab-scale precision or mechanically for industrial throughput, to separate casings, electrodes, and internal modules. Mechanical disintegration often incorporates cryogenic shredding or crushing to fragment battery structures, liberating embedded components such as anode/cathode materials and current collectors. This is followed by further comminution, which includes cryogenic grinding to detach electrode powders from aluminium or copper foils, thereby increasing liberation efficiency. Subsequent to comminution, the material undergoes classification, typically via sieving or air separation, to sort particles based on size and density. Fine particles (often < 0.25 mm) typically contain black mass rich in Li, Co, Ni, Mn, and graphite, while coarser fractions include metallic foils and casings. Additional separation techniques such as magnetic and eddy current separation are employed to isolate ferrous and non-ferrous metals, respectively. In some advanced setups, flotation or pneumatic separation is used to refine electrode fractions. To further detach electrode coatings from current collectors, thermal treatment or chemical dissolution methods are applied. Thermal treatment at approximately 450 °C is particularly effective in removing PVDF binders and residual electrolytes while avoiding contamination of active materials if optimised carefully. Alternatives such as organic solvents, deep eutectic solvents, and ionic liquids are also used to selectively dissolve binders [40–42].

The key output of the pretreatment stage is the black mass, a fine ground mixture of active materials from anodes and cathodes (Table 5). This fraction serves as the feedstock for hydrometallurgical or direct regeneration methods. Alongside black mass, several byproduct streams are recovered: Al foil (~5%–7% of total mass), Cu foil (~5%–10%), steel casings (~3%–5%), and plastic–organic residues (~30%–35%). Aluminium and copper are redirected to secondary metal recovery facilities, while steel is commonly recycled via standard remelting [19].

4. Conclusions and Discussion

The growing stream of WEE, driven by global demand for critical metals and concurrent digital and energy transitions, highlights both the urgency and the potential of eco-efficient recycling solutions. As demonstrated in this study, WEE such as printed circuit boards (WPCBs), hard disc drives (WHDs), and lithium-ion batteries (WLIBs) contains valuable metals ranging from copper and precious metals (Au, Ag, Pd) to critical elements (Co, Nd, Li) that can be recovered with substantial economic and environmental gains. Metals with the highest relative supply risk and ADP are Au, Pd, and Pt for WPCBs, Nd, Pr, Ce, Dy, Pd, and Au for WHDs, and Co, Li, and Ni for WLIBs. It should be noted that the estimated value of metals in 1 Mg of WPCBs could be almost USD 92000, while 1 Mg of WHDs could give recoverable metals worth approximately USD 9908, and 1 Mg of WLIBs could give a potential metal value exceeding USD 6351. To assess the profitability of recycling, investment and process costs must also be considered.

The annual mass of WPCBs, WHDs, and WLIBs collected per 1 million inhabitants of southern Poland is approximately 27.2, 13.6, and 45.3 Mg, respectively, and the estimated value of metals in them is approximately 2499, 135, and 288 thousand USD/year, respectively, indicating a large financial potential for companies involved in WEE recycling.

Physical recycling methods such as shredding, milling, electrostatic, magnetic, and gravity separation, complemented in the next step by hydrometallurgical or pyrometallurgical processes, enable the recovery of high-value metals at low process costs, lower energy consumption, and limited emissions of pollutants into the environment, compared to traditional mining. Such multi-step approaches are supported by continuous innovation in selective metal separation techniques and the application of new methods for the recovery of plastics and their use in the production of new consumer goods.

Furthermore, regional mass balance analysis demonstrates how improved collection and processing infrastructures can significantly increase the volume of recoverable metals. Converting end-of-life materials into secondary resources promotes regional economic development by creating jobs in recycling, metallurgy, and related industries and simultaneously mitigates the environmental damage associated with raw material extraction. In particular, recovering critical raw materials, such as cobalt and REEs, increases the resilience of the supply chain for industries dependent on these scarce inputs.

Beyond economic benefits, eco-efficient WEE recycling aligns with broader sustainability goals. Reduced landfill use, lower greenhouse gas emissions, and conservation of finite resources are key outcomes of developing recycling technologies aligned with the principles of a circular economy. Supporting research and development in key areas (e.g., advanced material sorting, optimised shredding, and selective separation) will further increase recycling efficiencies and contribute to global carbon neutrality targets.

It should also be noted that WEE, particularly WPCBs, contain metals at concentrations far exceeding those in geogenic ores. For example, based on the values in Table 1, the gold content in WPCBs is approximately 1000 g/t, whereas conventional gold ores often contain less than 5 g/t. Moreover, under optimised physical and metallurgical processing, the recovery efficiency of such metals from WEE can be comparable to—or even exceed—that of primary raw materials, which further supports the importance of WEE as a strategic secondary resource.

Implementing these solutions on a larger scale will be essential for sustainable resource management, resilient supply chains, and the protection of natural resources for future generations. By viewing WEE as a reservoir of high-value secondary raw materials rather than a waste problem, societies can capitalise on emerging market opportunities while adhering to responsible environmental stewardship in an increasingly resource-constrained world.

Beyond annual generation figures, it is essential to acknowledge that electronic waste is accumulating continuously at the global level. Most devices are not recycled promptly after use, resulting in growing stockpiles of unmanaged WEE. This long-term build-up not only increases the risk of environmental contamination from hazardous substances but also leads to the permanent loss of valuable secondary raw materials such as copper, gold, cobalt, and rare earth elements. These materials, once landfilled or processed using informal methods, are often no longer recoverable at scale.

In parallel, the accumulation of e-waste contributes to missed economic opportunities. According to recent global assessments, the embedded value of metals in e-waste exceeded USD 90 billion in 2022, much of which remains unrecovered. This inefficiency perpetuates reliance on primary extraction, increasing environmental degradation and geopolitical supply risks. Moreover, uncontrolled stockpiling of e-waste—whether in households, warehouses, or landfills—poses growing health threats in regions where informal dismantling and burning occur without proper protection.

Tackling the accumulation of e-waste calls for a coordinated and structural transformation across the entire management chain. Progress in recycling technology alone is insufficient without parallel advancements in collection efficiency, access to authorised treatment facilities, and the enforcement of extended producer responsibility (EPR) frameworks. Furthermore, effective policies must enable the gradual integration of informal recyclers into the formal economy by providing targeted support, regulatory clarity, and capacity-building programmes. Such a multidimensional approach is essential to mitigate the environmental and economic risks posed by unchecked e-waste growth and to move decisively toward a functioning circular economy aligned with long-term sustainability targets.

Although the global use of traditional HDDs and some lithium-ion battery types may have peaked, the accumulation of waste containing these components remains significant. Legacy equipment that has already been phased out still poses an environmental burden due to stockpiling or improper disposal. Moreover, the transition to SSDs and emerging battery chemistries such as silicon–carbon (Si-C) and lithium–sulphur (Li-S) technologies will soon create new streams of end-of-life products. Future studies should therefore address the recyclability and resource content of these innovative components to ensure preparedness for upcoming challenges in electronic waste management.

However, it must also be noted that metal price volatility—such as the decline observed in 2024—may discourage some companies from investing in recycling infrastructure, especially in cases where the recovered metals are of lower value. In this context, the consistently high market value of gold remains a major economic driver for recycling certain waste streams. Such risks underline the importance of developing integrated recovery systems that combine both high- and low-value materials to achieve overall profitability.

Author Contributions: Conceptualisation and formulation of theses, T.S.; formal analysis and conclusions, T.S.; methodology, T.S. and D.M.F.; analysis of metal price trends, T.S., P.F., U.K. and P.G.; analysis of content, price, and environmental, economic, and geopolitical importance of metals, T.S., P.F., U.K. and D.M.F.; analysis of economic and environmental potential of WEE recycling, T.S.; review of recycling research, P.F., U.K. and D.M.F.; investigation, T.S., P.F., U.K., D.M.F. and P.G.; writing—original draft preparation, T.S., P.F., U.K. and D.M.F.; writing—review and editing, T.S.; visualisation, T.S., P.F., U.K. and P.G. All authors have read and agreed to the published version of the manuscript.

Funding: The APC was funded by the KOMAG Institute of Mining Technology and the Silesian University of Technology.

Data Availability Statement: The data supporting the findings of this study are not publicly available due to commercial confidentiality agreements with the collaborating recycling companies.

Conflicts of Interest: The authors declare no conflict of interest.

References

1. Graedel, T.E.; Harper, E.M.; Nassar, N.T.; Nuss, P.; Reck, B.K. Criticality of Metals and Metalloids. *Proc. Natl. Acad. Sci. USA* **2015**, *112*, 4257–4262. [CrossRef] [PubMed]
2. United Nations Digital Library System. The Impact of Digital Technologies. Available online: https://www.un.org/en/un75/impact-digital-technologies (accessed on 28 March 2025).
3. International Energy Agency Low-Emissions Sources of Electricity-Analysis. Available online: https://www.iea.org/reports/low-emissions-sources-of-electricity (accessed on 28 March 2025).
4. Van Oers, L.; Guinée, J.B.; Heijungs, R. Abiotic Resource Depletion Potentials (ADPs) for Elements Revisited—Updating Ultimate Reserve Estimates and Introducing Time Series for Production Data. *Int. J. Life Cycle Assess.* **2020**, *25*, 294–308. [CrossRef]
5. Royal Society of Chemistry Periodic Table–Relative Supply Risk. Available online: https://periodic-table.rsc.org/ (accessed on 28 March 2025).
6. Franke, D.M.; Suponik, T.; Siudyga, T.; Kar, U. Evaluation of the Use of Flotation for the Separation of Ground Printed Circuit Boards. *Gospod. Surowcami Miner.-Miner. Resour. Manag.* **2022**, *38*, 171–188.
7. Mir, S.; Dhawan, N. Recycling of Metallic Values from Discarded Tantalum Capacitors. *Miner. Process. Extr. Metall.* **2023**, *132*, 87–98. [CrossRef]
8. Kaya, M. Recovery of Metals and Nonmetals from Waste Printed Circuit Boards (PCBs) by Physical Recycling Techniques. In *Energy Technology 2017*; Zhang, L., Drelich, J.W., Neelameggham, N.R., Guillen, D.P., Haque, N., Zhu, J., Sun, Z., Wang, T., Howarter, J.A., et al., Eds.; The Minerals, Metals & Materials Series; Springer International Publishing: Cham, Switzerland, 2017; pp. 433–451, ISBN 978-3-319-52191-6.
9. Macrotrens LLC Macrotrends | The Long Term Perspective on Markets. Available online: https://macrotrends.net (accessed on 28 March 2025).
10. IndexMundi LCC Commodity Prices. Available online: https://www.indexmundi.com/commodities/ (accessed on 28 March 2025).
11. London Metal Exchange, T.L.M. Metals. Available online: https://www.lme.com/ (accessed on 18 March 2025).
12. Manhart, A.; Buchert, M.; Degreif, S.; Mehlhart, G.; Meinel, J. *Recycling of Hard Disk Drives–Analysing the Optimal Dismantling Depth for Recyclers in Developing Countries and Emerging Economies*; Oko-Institut: Freiburg, Germany, 2015.
13. Habib, K.; Parajuly, K.; Wenzel, H. Tracking the Flow of Resources in Electronic Waste-the Case of End-of-Life Computer Hard Disk Drives. *Environ. Sci. Technol.* **2015**, *49*, 12441–12449. [CrossRef] [PubMed]
14. Ueberschaar, M.; Rotter, V.S. Enabling the Recycling of Rare Earth Elements through Product Design and Trend Analyses of Hard Disk Drives. *J. Mater. Cycles Waste Manag.* **2015**, *17*, 266–281. [CrossRef]
15. Trading Economics LCC Neodymium. Available online: https://tradingeconomics.com/commodity/neodymium (accessed on 9 April 2025).
16. IDC Global Worldwide Hard Disk Drive Forecast Update, 2023–2027. Available online: https://www.idc.com/getdoc.jsp?containerId=US51423423 (accessed on 28 March 2025).
17. Trendfocus Inc. The Data Storage Industry's Most Trusted Market Intelligence. *Pioneering Analysis in SSDs, HDDs, NAND in Cloud, Enterprise, and Cloud Markets.* Available online: https://trendfocus.com/ (accessed on 28 March 2025).
18. Kaya, M. State-of-the-Art Lithium-Ion Battery Recycling Technologies. *Circ. Econ.* **2022**, *1*, 100015. [CrossRef]
19. An, L. (Ed.) *Recycling of Spent Lithium-Ion Batteries: Processing Methods and Environmental Impacts*; Springer International Publishing: Cham, Switzerland, 2019; ISBN 978-3-030-31833-8.
20. Horeh, N.B.; Mousavi, S.M.; Shojaosadati, S.A. Bioleaching of Valuable Metals from Spent Lithium-Ion Mobile Phone Batteries Using Aspergillus Niger. *J. Power Sources* **2016**, *320*, 257–266. [CrossRef]
21. World Bank Group Commodity Prices Revised Down as Global Growth Weakens and Supplies Remain Ample. Available online: https://www.worldbank.org/en/news/press-release/2019/10/29/commodity-prices-revised-down-as-global-growth-weakens-and-supplies-remain-ample (accessed on 28 March 2025).
22. Statista Ltd. Mine Production of Rare Earths Worldwide in 2024, by Leading Country. Available online: https://www.statista.com/statistics/268011/top-countries-in-rare-earth-mine-production/ (accessed on 10 April 2025).
23. Cucchiella, F.; D'Adamo, I.; Lenny Koh, S.C.; Rosa, P. Recycling of WEEEs: An Economic Assessment of Present and Future e-Waste Streams. *Renew. Sustain. Energy Rev.* **2015**, *51*, 263–272. [CrossRef]
24. De Oliveira Neto, G.C.; De Jesus Cardoso Correia, A.; Schroeder, A.M. Economic and Environmental Assessment of Recycling and Reuse of Electronic Waste: Multiple Case Studies in Brazil and Switzerland. *Resour. Conserv. Recycl.* **2017**, *127*, 42–55. [CrossRef]

25. Luo, X.; Ding, N.; Yang, J.; Lu, B.; Ma, J. Potential Environmental Benefits Assessment of Recycling Based on Multi-LCA and SFA. *J. Clean. Prod.* **2024**, *457*, 142370. [CrossRef]
26. E-Waste Monitor. *The Global E-Waste Monitor 2024*; E-Waste Monitor: Bonn, Germany, 2024.
27. Valdez, M. *The Biggest E-Waste Facts of 2024*; Reboot Tech LLC: ITAD & eWaste Recycling: Montclair, CA, USA, 2024.
28. Franke, D.; Suponik, T.; Nuckowski, P.; Dubaj, J. Evaluation of the Efficiency of Metal Recovery from Printed Circuit Boards Using Gravity Processes. *Physicochem. Probl. Miner. Process.* **2021**, *57*, 63–77. [CrossRef]
29. Hyra, K.; Nuckowski, P.M.; Willner, J.; Suponik, T.; Franke, D.; Pawlyta, M.; Matus, K.; Kwaśny, W. Morphology, Phase and Chemical Analysis of Leachate after Bioleaching Metals from Printed Circuit Boards. *Materials* **2022**, *15*, 4373. [CrossRef]
30. Jin, H.; Afiuny, P.; Dove, S.; Furlan, G.; Zakotnik, M.; Yih, Y.; Sutherland, J.W. Life Cycle Assessment of Neodymium-Iron-Boron Magnet-to-Magnet Recycling for Electric Vehicle Motors. *Environ. Sci. Technol.* **2018**, *52*, 3796–3802. [CrossRef] [PubMed]
31. Klemettinen, A.; Adamski, Z.; Chojnacka, I.; Leśniewicz, A.; Rycerz, L. Recovery of Rare Earth Elements from the Leaching Solutions of Spent NdFeB Permanent Magnets by Selective Precipitation of Rare Earth Oxalates. *Minerals* **2023**, *13*, 846. [CrossRef]
32. European Aluminium Environmental Profile Report 2018-Executive Summary. Available online: https://european-aluminium.eu/news_events/environmental-profile-report-2018-executive-summary/ (accessed on 10 April 2025).
33. Park, I.; Ito, M.; Jeon, S.; Tabelin, C.B.; Phengsaart, T.; Silwamba, M.; Hiroyoshi, N. A Novel Recycling Route for Aluminum Alloys: Synthesis of Fe/Al Bimetallic Materials and Magnetic Separation. *Miner. Eng.* **2023**, *201*, 108202. [CrossRef]
34. Sun, B.; Sun, N.; Wang, L.; Zhang, S.; Guo, Z. A New Process for Efficient Purification of 6061 Aluminum Alloy Scrap under Semi-Solid and Super-Gravity Conditions. *J. Clean. Prod.* **2023**, *402*, 136825. [CrossRef]
35. Friebe, P. Development of a Prototype Shredder for WEEE Equipped with NdFeB Magnets. *Min. Mach.* **2023**, *41*, 143–157. [CrossRef]
36. Friebe, P.; Suponik, T.; Nuckowski, P.M.; Matusiak, P.; Kowol, D. A Method of Recovering Metals from Used Hard Drives. Poland Patent P.452083, 19 May 2025.
37. Al-Thyabat, S.; Nakamura, T.; Shibata, E.; Iizuka, A. Adaptation of Minerals Processing Operations for Lithium-Ion (LiBs) and Nickel Metal Hydride (NiMH) Batteries Recycling: Critical Review. *Miner. Eng.* **2013**, *45*, 4–17. [CrossRef]
38. Zhao, Y.; Yuan, X.; Jiang, L.; Wen, J.; Wang, H.; Guan, R.; Zhang, J.; Zeng, G. Regeneration and Reutilization of Cathode Materials from Spent Lithium-Ion Batteries. *Chem. Eng. J.* **2020**, *383*, 123089. [CrossRef]
39. Kim, S.; Bang, J.; Yoo, J.; Shin, Y.; Bae, J.; Jeong, J.; Kim, K.; Dong, P.; Kwon, K. A Comprehensive Review on the Pretreatment Process in Lithium-Ion Battery Recycling. *J. Clean. Prod.* **2021**, *294*, 126329. [CrossRef]
40. Kar, U.; Nili, S.; Mends, E.; Vahidi, E.; Chu, P. A Review and Environmental Impact Analysis on the Current State of Froth Flotation on Recycling of E-Wastes. *Resour. Conserv. Recycl.* **2025**, *212*, 107967. [CrossRef]
41. Gao, T.; Dai, T.; Fan, N.; Han, Z.; Gao, X. Comprehensive Review and Comparison on Pretreatment of Spent Lithium-Ion Battery. *J. Environ. Manag.* **2024**, *363*, 121314. [CrossRef] [PubMed]
42. Sommerville, R.; Shaw-Stewart, J.; Goodship, V.; Rowson, N.; Kendrick, E. A Review of Physical Processes Used in the Safe Recycling of Lithium Ion Batteries. *Sustain. Mater. Technol.* **2020**, *25*, e00197. [CrossRef]

minerals

MDPI

Article

Comparative Study of the Effect of Particle Size on Flotation Kinetics of Raw and Waste Coal

Jovica Sokolović *, Ivana Đolović, Dejan Tanikić, Zoran Štirbanović and Ivana Ilić

University of Belgrade, Technical Faculty in Bor, Vojske Jugoslavije 12, 19210 Bor, Serbia; idjolovic@tfbor.bg.ac.rs (I.Đ.); dtanikic@tfbor.bg.ac.rs (D.T.); zstirbanovic@tfbor.bg.ac.rs (Z.Š.); iilic@tfbor.bg.ac.rs (I.I.)

* Correspondence: jsokolovic@tfbor.bg.ac.rs

Abstract

This study examines the influence of particle size on the flotation kinetics parameters of both raw and waste fine coal originating from the anthracite mine "Vrška Čuka", Serbia. Flotation kinetics modeling was performed using MATLAB for nonlinear regression analysis, based on coal flotation test data. The correlation between total combustible recovery and flotation time was determined using the following models: Classical, Klimpel, Kelsall, Modified Kelsall, and Fully Mixed. The coefficients of determination range from 0.9724 (the Klimpel model) to 1 (the modified Kelsall model) for raw coal and from 0.8609 (the Klimpel model) to 0.9981 (the modified Kelsall model) for waste coal. Although both the Classical and Modified Kelsall models demonstrated a good correlation with the experimental data, the Modified Kelsall model provided a slightly better fit. The maximum values of the flotation rate constant (k) for both coals were obtained for the particle size-class (−0.1 + 0.053) mm for the Classical model and (−0.2 + 0.1) mm for the modified Kelsall model. The relation between flotation kinetics constant (k) and average particle size value (d_{sr}) was estimated for the Classical model and the modified Kelsall model. It was observed that the flotation kinetics constant (k) for coal particle size could be predicted satisfactorily.

Keywords: coal; particle size; flotation kinetics; classical model; modified Kelsall model

1. Introduction

Coal flotation is a very complex physicochemical process and difficult to control. Coal flotation plays an important role in coal processing and is a commonly used method in coal processing technologies for the recovery of fine coal below 0.5 mm [1] and in some cases even below 1 mm [2].

The coal flotation theory and parameters affecting the coal flotation performance are explained in detail by Polat [1] and Laskowski [3]. Generally, a wide range of parameters affects coal flotation [1]. The most critical parameter affecting the stability of bubble-particle aggregates is the particles' size [4]. Particle size has an important role in coal flotation as well as flotation kinetics.

A number of kinetic models have been developed and tested for the coal flotation process. A comprehensive review on flotation kinetics was presented by Lynch et al. [5], Polat and Chander [6], and Drzymala [7]. A review of the available literature indicates that coal flotation typically follows first-order kinetics. In general, the best-fit kinetic model varies for different coal types and flotation conditions.

The relationship between particle size and flotation performance is complex, as demonstrated by numerous studies showing that variations in coal particle size distribution significantly affect flotation kinetics. Numerous studies have been carried out to investigate the influence of particle size on the coal flotation kinetics [8–32].

Sokolović and Mišković [8] provided a comprehensive review of flotation kinetics models for coal presented in the literature, summarizing key findings from selected studies that examined the influence of particle size on flotation recovery, rate constants, and quality, while also emphasizing major developments and results achieved in this field over the past several decades.

Li et al. [17] studied the flotation kinetics and separation selectivity of three narrow coal size fractions. The values of R and k were determined using MATLAB R2010a software. They assert that during coal flotation, the organic component floats according to the first-order kinetics, while the inorganic component floats according to the second-order kinetics. The flotation rate constant of both combustible matter and ash-forming minerals increases with the particle size.

Ni et al. [19] used six kinetic models to coal flotation and found that the first-order kinetic model with rectangular distribution best described the process of the rougher and cleaner flotation of a bituminous coal.

Bu et al. [20] reported that the representative flotation kinetic models of 375, 188, 100, and 37 μm (average particle sizes) coal particles corresponded to the first order with rectangular distribution, non-integral order, non-integral order, and first order with rectangular distribution, respectively. Similarly, Bu et al. [22] concluded that the first-order model best fits conventional flotation of −74 μm coal fines, while more complex models are suited for carrier flotation.

Shahbazi and Chelgani [23] developed computational models correlating particle properties and hydrodynamic conditions with flotation performance, achieving high predictive accuracy (R^2 = 0.85–0.99) for flotation rate and recovery.

Norori-McCormac et al. [24] demonstrated that particle size distribution substantially affects froth stability. Specifically, intermediate distributions ($d_{80} \approx 103.5$ μm) generate more stable froths at lower air flow rates, while finer particles ($d_{80} \approx 89.6$ μm) favor higher recovery at increased air rates.

The introduction of novel flotation technologies, such as oily bubble flotation [21,25] and nanobubble-assisted flotation [26], has further enhanced the recovery and kinetics of medium and fine coal particles. These technologies deliver higher flotation rate constants and improved combustible recovery compared to conventional air bubble flotation, emphasizing the role of particle size in flotation mechanism efficiency.

Studies by Sahoo et al. [27] and Bahrami et al. [28] confirmed that intermediate particle size fractions achieve the highest flotation rates and combustible recoveries in both bituminous and high-ash coals. In both cases, the first-order kinetic model, combined with a rectangular floatability distribution, provided the best fit to experimental data, validating the model's applicability across coal types and particle size ranges.

Şahbaz and Demir [29] revealed that frother type and particle size jointly affect flotation kinetics in the Jameson cell. Their study found that aliphatic alcohol-based frothers improve the recovery of fine particles, whereas polypropylene glycol-based frothers are more effective for coarser fractions. According to Karaca et al. [30], the flotation rate constant increased significantly for medium and coarse particles in the Jameson cell compared to mechanical flotation cells.

Recent studies by Lu et al. [31] and Kazemi et al. [32] further emphasized the crucial role of particle size distribution and liberation characteristics. Lu et al. [31] demonstrated that medium-sized coal slime fractions exhibit superior hydrophobicity and flotation kinetics, while Kazemi et al. [32] showed that recovery efficiencies vary significantly between laboratory and industrial scales, influenced by particle size and liberation degree.

In summary, the flotation rate constant is strongly dependent on coal particle size. More recent studies have shown that the highest flotation rates are achieved within an intermediate particle size range, while significantly lower rates are observed for both fine and coarse particles. These trends are commonly attributed to the low collision probability between fine particles and bubbles, as well as the high likelihood of detachment of coarse particles from bubbles.

The relation between the flotation rate constant and cumulative recovery with particle size was found to be nonlinear [15]. Studies showed that the first-order kinetic model with a rectangular distribution of floatability gave the best fit to the flotation experimental data for coal with particle sizes between 37 and 375 μm [8]. Furthermore, the nonintegral-order equation fit the test data of fine coal with average particle sizes between 188 and 100 μm [15].

Understanding the relationship between particle size and flotation kinetics is crucial for optimizing the performance of the coal flotation process. Despite previous studies, inconsistencies remain in reports on the optimal particle sizes for maximum recovery and rate constants, partly due to variations in coal rank, surface chemistry, and experimental conditions. On the other hand, coal flotation kinetics are insufficiently studied on waste coals, i.e., coals that have been exposed to oxidation due to atmospheric influences.

Natural weathering processes can make coal surfaces more hydrophilic due to an increase in the content of hydrophilic functional groups (C-O, C=O, and COOH) and a decrease in the content of hydrophobic functional groups (C-C and C-H) on the coal surface [33]. These functional groups significantly reduce the wettability and hydrophobicity of coal [34] and, thus, negatively affect its flotation [35,36]. Weathering can affect the parameters used to model flotation kinetics, such as the flotation rate constant, thus reducing the flotation recovery of coal [37]. Sokolovic et al. [37] considered a classical first-order model and the modified Kelsall model for modeling raw and waste anthracite coal flotation. The application of a nonlinear model fitting approach showed that both models exhibited good correlation with the experimental data, with the Modified Kelsall model providing the best fit.

It is widely recognized that there is a strong relationship between particle size and the flotation kinetics. On the other hand, the flotation kinetics of waste coal have been little investigated. This paper investigates the effect of particle size on the flotation kinetics parameters of raw and waste coal from the anthracite mine "Vrška Čuka", Serbia. The aim of this study is to systematically investigate the effect of particle size on the flotation kinetics of raw and waste coal, using well-defined particle size-classes and controlled flotation conditions. The results are expected to provide insights into the rate-limiting steps for different particle size ranges and guide process optimization for enhanced recovery and product quality. Understanding these interactions allows for more efficient control over flotation processes and better separation efficiency.

2. Materials and Methods

2.1. Materials

Coal samples of about 250 kg were collected from the anthracite coal mine "Vrška Čuka" in Serbia. The first representative samples were collected from raw coal, and the second from the coal waste ponds. Raw coal (RC) was sampled as feed from the "BSRI-1200" dense-medium separator, with a particle size range of 0–25 mm. Waste coal (WC) was sampled from the coal waste ponds, with a particle size range of 0–1 mm. The raw coal sample was screened at 1 mm for particle size analysis and flotation kinetics tests. The collected samples were subsampled using the coning and quartering method to obtain a representative sample for particle size-class analysis and laboratory flotation tests.

2.2. Methods

2.2.1. Particle Size-Class Analysis

The raw and waste coal samples were sieved to determine the particle size-class distribution. Wet sieving on sieves 0.5, 0.2, 0.1, and 0.053 mm was applied for determining the particle size distribution of both coal samples.

Laboratory sieve shaker Retsch AS200 was used with an amplitude of 2 mm for 25 min. After drying, the ash content was determined according to the ASTM D3174 [38] standard from 2014 at 815 °C.

The particle size distribution-wise ash content is shown in Table 1.

Table 1. Particle size distribution and ash contents of raw coal (RC) and waste coal (WC) samples.

Particle Size Fraction d (mm)	Raw Coal (RC)				Waste Coal (WC)			
	Internal Values		Cumulative Passing Value		Internal Values		Cumulative Passing Value	
	Weight (%)	Ash (%)	Weight (%)	Ash (%)	Weight (%)	Ash (%)	Weight (%)	Ash (%)
−1 + 0.5	23.52	20.98	23.52	20.98	23.96	31.84	23.96	31.84
−0.5 + 0.2	21.04	14.54	44.56	17.94	28.31	29.60	52.27	30.63
−0.2 + 0.1	13.94	16.88	58.50	17.69	14.91	35.73	67.18	31.76
−0.1 + 0.053	15.31	17.22	73.81	17.59	10.05	42.20	77.23	33.12
−0.053 + 0	26.19	22.31	100.00	18.83	22.77	53.23	100.00	37.70

Based on the results presented in Table 1, it can be concluded that raw and waste coal show a similar mass distribution by size-class. By comparing the particle size distribution and ash contents by size-classes, it can be seen that a significant partition of both tested coal samples lies in the (−1 + 0.2) mm particle size range. Additionally, the presence of particles coarser than 0.5 mm was about 24% for both coals. The lowest ash content in both tested coal samples was observed in the (−0.5 + 0.2) mm particle size-class, amounting to 14.54% for raw coal and 29.60% for waste coal. It was noted that fine particle size-classes below 0.053 mm are 26.19% for raw coal and 22.77% for waste coal. Furthermore, higher ash content was observed in waste coal by particle size-classes. The average ash content in the waste coal sample was about 37%, compared to about 19% in the raw coal.

Higher ash content in waste coal may result from coal oxidation during weathering, a process whereby chemical changes occur in the coal's structure upon prolonged exposure to atmospheric factors (such as oxygen, water, moisture, and other elements), particularly in coal storage ponds. This oxidation process induces the formation of oxygen-containing functional groups on the coal surface, leading to an increase in ash content and a subsequent reduction in flotation efficiency.

2.2.2. Proximate and Ultimate Analyses

The proximate and ultimate analyses of raw and waste coal were conducted at the Institute of Mining and Metallurgy (MMI) Bor, utilizing standard ASTM procedures to determine the coal's composition. The CHNOS analysis (carbon, hydrogen, nitrogen, and sulfur) was performed using a LECO CHN628 elemental analyzer. The results of proximate and ultimate analyses of the raw and waste samples are given in Table 2.

Table 2. Proximate and ultimate analyses of the raw and waste coal.

	Raw Coal	Waste Coal	Method
Proximate analysis			
Moisture (%)	4.34	7.69	ISO 589 [39]
Ash (%)	18.53	36.31	ASTM D3174 [38]
Volatile matter (%)	6.43	8.94	ASTM D3175 [40]
Fixed carbon (%)	70.70	47.06	R
Ultimate analysis			
C (%)	72.69	56.00	ASTM D5373 [41]
H (%)	3.32	2.72	ASTM D5373 [41]
N (%)	0.65	0.55	ASTM D5373 [41]
S (%)	2.29	2.31	ASTM D4239 [42]
Calorific Value (kJ/kg)	29,065	21,954	ASTM D5865 [43]

Table 2 shows that the highest carbon, hydrogen, and nitrogen content were observed in the raw coal. Also, the results of proximate analysis confirm the ash content values shown in Table 1. The carbon content was found to be higher (72.69%) in the raw coal, while the ash content was found to be higher (36.31%) in the waste coal. The results of proximate and ultimate analyses of the raw and waste coal show a reduction in C/H ratio (obtained from ultimate analysis) for about 1.30% in the waste coal, indicating the sensitivity of the coal surface to oxidation (weathering).

2.2.3. Flotation Tests

A laboratory flotation machine, Denver D-12, was used for flotation kinetics tests with the impeller speed of 1250 rpm. All coal flotation tests were conducted with feed slurries at 10% solids (by weight). The pH was adjusted to 7.5. The pH value of 7.5 was selected based on extensive experimental research [37] and corresponds to the isoelectric point of coal determined through zeta potential measurements, making it the optimal pH for flotation. Kerosene (1000 g/t), tannic acid (200 g/t), sodium silicate (200 g/t), and pine oil (45 g/t) were used as reagents in this study. The kerosene collector dosage of 1 kg/t was identified as the optimal value through flotation experiments [36] and reflects the industrial parameters used in the flotation process at the Vrška Čuka anthracite mine.

In each flotation kinetics test, the pulp was first agitated in the flotation cell for 3 min. Agitation was performed for 2 min, after which kerosene was added and the slurry was mixed for an additional 0.5 min. Subsequently, tannic acid, sodium silicate, and pine oil were introduced, followed by further stirring for 0.5 min. Air was introduced after the conditioning stage, and flotation concentrates were collected at 30, 90, 150, and 360 s. Concentrates and tailings obtained from froth flotation were processed by filtration, drying, and weighing, followed by ash content analysis for each sample. Upon completion of the flotation kinetics experiments, concentrates were subjected to dry sieving. Size-wise ash analysis for the five particle size-classes, 0.5, 0.2, 0.1, and 0.053 mm, was conducted for froth concentrates and the tailings. The sieve analysis was performed to determine the particle size distribution of the flotation products, which is critical for understanding the efficiency and selectivity of the flotation process. Based on the obtained results, the combustible recovery (R) of non-ash materials was calculated by:

$$R = Y\left(\frac{100 - Ac}{100 - Af}\right) \quad (1)$$

where Y is the percentage yield of the concentrate, and Ac and Af are the percentage ash contents in concentrates and feed materials, respectively [12].

2.2.4. Flotation Kinetics

In this study, six different kinetic flotation models, Classical model (Zuniga, 1935) [44], Klimpel model (Klimpel, 1980) [45], Kelsall model (Kelsall, 1961) [46], Modified Kelsall model (Jowett, 1974) [47], and Fully mixed model (Imaizumi and Inoue, 1965) [48], were used to analyze batch flotation results and to optimize the flotation parameters using kinetics models.

1. Classical model

A classical first-order kinetics model, introduced by Zuniga (1935) [44], given in Equation (2), is used to obtain the parameters of k and R_∞ through fitting the flotation test data.

$$R = R_\infty\left(1 - e^{-kt}\right) \quad (2)$$

where R is the recovery [%] (dependent variable), R_∞ is the ultimate (maximum) recovery of non-ash materials [%] (parameter 1), k is the flotation rate constant [min^{-1}] (parameter 2, and t is the cumulative flotation time [min] (independent variable).

The classical first-order kinetics model was chosen to show the trends observed in the data due to its simplicity and the relatively good fit it provided.

2. Klimpel model

Klimpel model [45] used two parameters to describe flotation kinetics:

$$R = R_\infty\left[1 - \frac{1}{kt}\left(1 - e^{-kt}\right)\right] \quad (3)$$

where R_∞ is the ultimate (maximum) recovery of non-ash materials (%) (parameter 1), and k is the modified flotation rate constant (min^{-1}) (parameter 2).

3. Kelsall model

Kelsall model [46] describes the flotation kinetics by considering the presence of two types of particles: fast-floating particles, with a rate constant k_f, and slow-floating particles, with a rate constant k_s. Additionally, the model introduces the parameter φ, which represents the fraction of slow-floating particles with the slow rate constant. Equation (4), which describes this model, can be expressed as follows:

$$R = (1 - \varphi)\left(1 - e^{-k_f t}\right) + \varphi\left(1 - e^{-k_s t}\right) \quad (4)$$

In this model, the ultimate recovery (R_∞) is not included. This model provides a slightly better description of experimental results compared to other models due to the additional parameter φ.

4. Modified Kelsall model

A modified Kelsall model [47] provides a slight improvement over the previous one, Equation (4), by introducing the value for ultimate recovery (R_∞), and is represented by Equation (5):

$$R = R_\infty\left[(1 - \varphi)\left(1 - e^{-k_f t}\right) + \varphi\left(1 - e^{-k_s t}\right)\right] \quad (5)$$

where k_f is the fast flotation rate constant (min^{-1}), k_s is the slow flotation rate constant (min^{-1}), and φ is the fraction of flotation components with the slow rate constant.

5. Fully mixed model

The fully mixed model [48], as the classical model, uses R_∞ and k to describe flotation kinetics:

$$R = R_\infty\left[1 - \left(\frac{1}{1+\frac{t}{k}}\right)\right] \tag{6}$$

The MATLAB tool was used to fit the first-order kinetics model to the flotation test results to obtain R_∞ and k values, i.e., k (first-order rate constant), k_f (fast flotation rate constant), and k_s (slow flotation rate constant) [49].

The coefficient of determination (r^2) was used to evaluate the accuracy of the model fit, and if it was higher than 0.8, Equations (2)–(6) can be applied. Results are presented and discussed in the next section.

3. Results and Discussion

3.1. Particle Size Analysis of the Coal Flotation Products

Figures 1 and 2 present a particle size-class analysis of the flotation products of raw and waste coal.

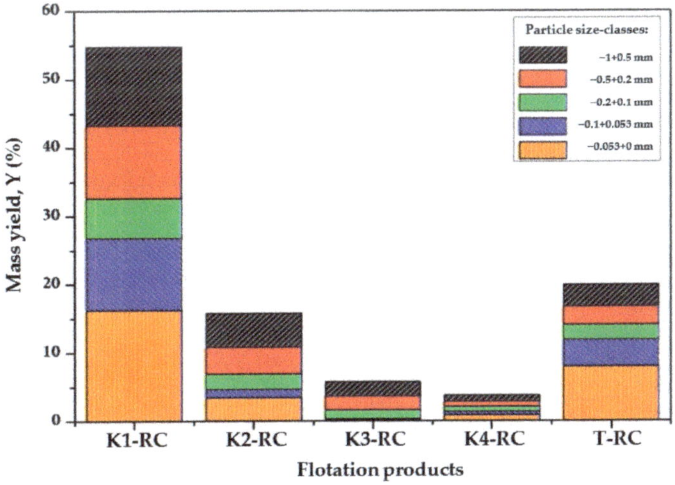

Figure 1. Particle size-class analysis of the flotation products of raw coal.

From Figure 1, it can be seen that the dominant yield in the first minutes of raw coal flotation is composed of the coarsest and smallest coal particles. The significant presence of particle size-classes (−1 + 0.5) mm and (−0.5 + 0.2) mm in the coal flotation products obtained after 2.5 and 6 min of flotation, respectively, indicates that particles within these size-classes have the highest flotation rates. Compared to other particle size-classes, the cumulative shares of these particle size-classes in the final concentrates are 24.45% and 21.70%, respectively.

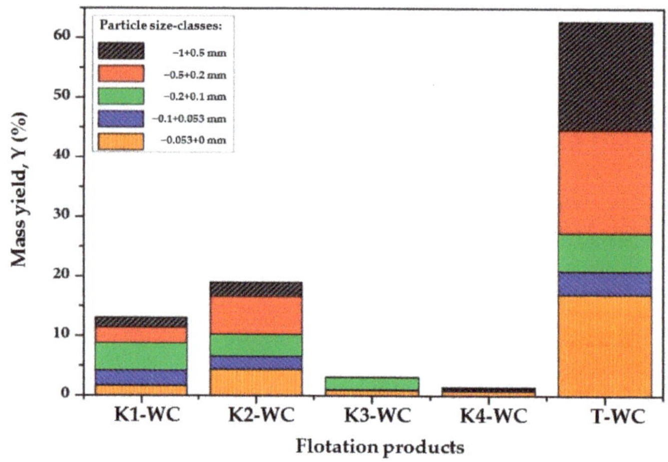

Figure 2. Particle size-class analysis of the flotation products of waste coal.

An analysis of the particle size-class distribution of the concentrate obtained from the flotation of waste coal (Figure 2) shows that coal particles from the size-classes (−0.5 + 0.2) mm and (−0.2 + 0.1) mm exhibit the highest flotation probability within the first few minutes. Notably, the (−0.5 + 0.2) mm size fraction also has the lowest ash content, which is advantageous for coal flotation. It is evident that the optimal size shifts towards intermediate and finer particle size-classes, specifically the (−0.2 + 0.1) mm and (−0.1 + 0.053) mm classes. Compared to raw coal, the mass yield of these size-classes in the final concentrates is higher, reaching approximately 28%. In contrast, the cumulative yield of the coarsest particles larger than 0.5 mm is 9.25% and 21.05% for the finest class (−0.053 + 0 mm).

3.2. Effect of Coal Particle Size on Flotation Recovery

The effect of coal particle size on the flotation recovery of raw and waste coal is presented in Tables 3 and 4, respectively.

Table 3. Flotation kinetics of different particle size-classes of raw coal (RC).

Flotation Products	Flotation Time [s]	Cumulative Recovery [%]				
		−1 + 0.5 mm	−0.5 + 0.2 mm	−0.2 + 0.1 mm	−0.1 + 0.053 mm	−0.053 + 0 mm
K1-RC	30	55.88	58.85	52.63	72.55	63.49
K2-RC	90	81.54	81.77	75.08	81.29	77.43
K3-RC	150	90.30	90.97	85.76	82.11	78.20
K4-RC	360	93.37	93.74	90.68	82.28	81.29
T-RC		6.63	6.26	9.32	17.72	18.71

The cumulative recoveries for the individual particle size-classes were plotted as a function of flotation time. The curves of flotation kinetics of different particle size-classes of raw coal (RC) and waste coal (WC) are shown in Figures 3 and 4.

Table 4. Flotation kinetics of different particle size-classes of waste coal (WC).

Flotation Products	Flotation Time [s]	Cumulative Recovery [%]				
		−1 + 0.5 mm	−0.5 + 0.2 mm	−0.2 + 0.1 mm	−0.1 + 0.053 mm	−0.053 + 0 mm
K1-WC	30	6.46	18.79	37.25	38.41	9.59
K2-WC	90	21.74	48.62	63.18	66.65	33.42
K3-WC	150	22.10	48.95	74.88	68.14	38.39
K4-WC	360	22.23	49.62	75.86	71.49	42.50
T-WC		77.67	50.38	24.14	28.51	57.50

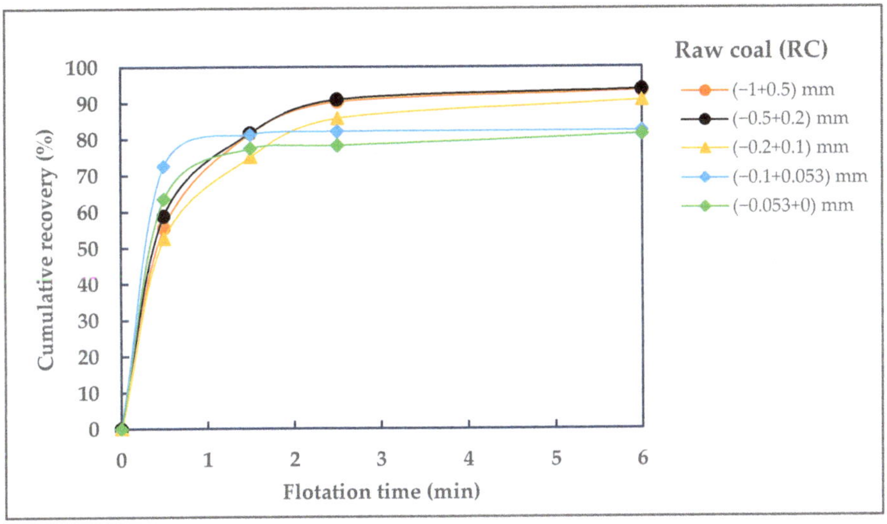

Figure 3. Cumulative recovery of different particle size-classes of raw coal (RC) against flotation time.

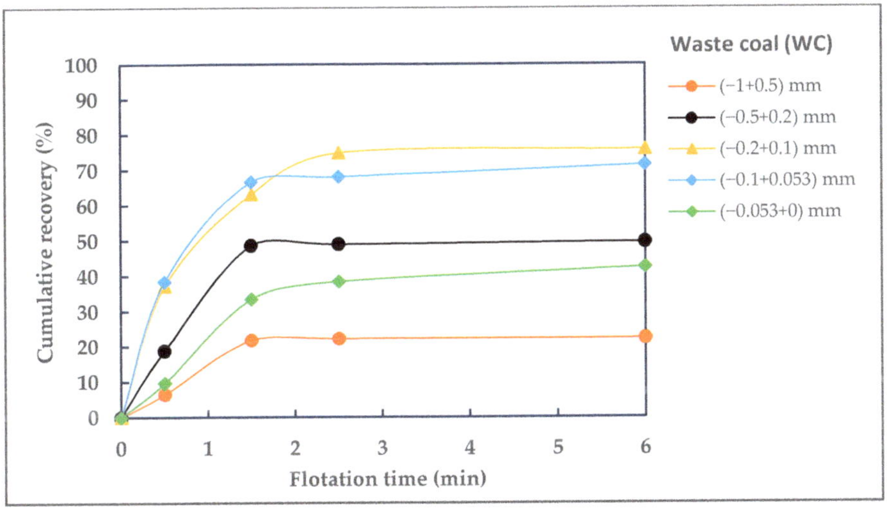

Figure 4. Cumulative recovery of different particle size-classes of waste coal (WC) against flotation time.

The good flotation kinetics and achieved flotation recovery (over 80%) indicate the successful flotation of all coal particle sizes. The best flotation kinetics are observed for the particle size fraction (−0.5 + 0.2) mm.

However, the results show that fine, hydrophobic coal particles (below 0.1 mm) float at the beginning of the process, while the flotation rate of coarser particles (above 0.5 mm) increases after approximately 1.5 min, following the flotation of the finer particles.

For waste coal, different results were obtained. As shown in Figure 4, the difference in flotation recoveries is particularly pronounced for the coarsest and finest coal particles. Compared to raw coal, the flotation recovery of particles larger than 0.5 mm is significantly lower. The recovery of 22.23% for particles larger than 0.5 mm confirms this conclusion. The most efficient flotation of waste coal is observed in the size fractions (−0.2 + 0.1) mm and (−0.1 + 0.053) mm, which, from the point of flotation, represent the optimal particle size range. Compared to raw coal, the recovery of these fractions is lower, ranging from 71.49% to 75.86%. Furthermore, waste coal particles below 0.053 mm are more difficult to float compared to raw coal. This reduced floatability is primarily due to their higher hydrophilicity and ash content, both of which adversely affect flotation efficiency. The recovery for the finest particle size-class (−0.053 + 0) mm is 42.50%.

3.3. *Effect of Coal Particle Size on Flotation Kinetics Parameters*

Using the MATLAB tool for modeling first-order flotation kinetics, the values of the flotation rate constants and other flotation parameters were determined for five flotation kinetics models.

The obtained values of the flotation kinetics parameters for various particle size-classes of raw coal (RC) and waste coal (WC) samples are shown in Tables 5 and 6.

Table 5. Kinetics parameters generated by the first-order classical model for different particle size-classes of raw coal (RC) sample (r^2—the coefficient of determination).

Flotation Kinetics Model	Flotation Kinetics Parameters	Particle Size-Classes				
		−1 + 0.5 mm	−0.5 + 0.2 mm	−0.2 + 0.1 mm	−0.1 + 0.053 mm	−0.053 + 0 mm
Classical	R_∞(%)	93.37	93.74	90.68	82.28	81.29
	k (min^{-1})	1.6848	1.8084	1.5017	4.2624	2.9839
	r^2	0.9996	0.9942	0.9904	0.9998	0.9980
Klimpel	R_∞(%)	98.45	97.999	94.518	84.117	82.926
	k (min^{-1})	4.0118	4.4288	3.7273	14.9073	8.4928
	r^2	0.9953	0.9921	0.9888	0.9724	0.9908
Kelsall	k_f (min^{-1})	2.1056	2.3986	2.2702	4.417	3.5063
	k_s (min^{-1})	0.1728	0.2132	0.1940	0.0087	0.0391
	φ	0.1711	0.1920	0.2698	0.1760	0.2374
	r^2	0.9988	0.9981	0.9976	0.9996	0.99998
Modified Kelsall	R_∞(%)	93.97	94.52	91.23	82.36	81.30
	k_f (min^{-1})	3.9956	5.9	25	4.7705	3.5063
	k_s (min^{-1})	1	1	0.9119	1	0.0391
	φ	0.5730	0.5865	0.6696	0.0526	3.3109×10^{-12}
	r^2	0.9999	0.9997	0.9998	1	0.99998
Fully Mixed	R_∞(%)	100.00	100.00	98.21	84.36	83.79
	k (min^{-1})	0.3652	0.3351	0.4326	0.07814	0.1553
	r^2	0.9977	0.9985	0.9988	0.9995	0.9993

Table 6. Kinetics parameters generated by the first-order classical model for different particle size-classes of waste coal (WC) sample (r^2—the coefficient of determination).

Flotation Kinetics Model	Flotation Kinetics Parameters	Particle Size-Classes				
		−1 + 0.5 mm	−0.5 + 0.2 mm	−0.2 + 0.1 mm	−0.1 + 0.053 mm	−0.053 + 0 mm
Classical	R_∞(%)	23.55	51.76	76.26	71.70	44.06
	k (min^{-1})	1.0584	1.2056	1.2877	1.4557	0.7786
	r^2	0.9555	0.9746	0.9980	0.9960	0.9789
Klimpel	R_∞(%)	26.33	57.29	83.35	77.32	50.08
	k (min^{-1})	2.0665	2.4325	2.7660	3.4611	1.4777
	r^2	0.8609	0.8838	0.9797	0.9638	0.9468
Kelsall	k_f (min^{-1})	1.0584	1.2056	1.2877	1.6158	0.7802
	k_s (min^{-1})	2.2216×10^{-14}	2.2207×10^{-14}	2.2204×10^{-14}	0.0025	2.2204×10^{-14}
	φ	0.7645	0.4824	0.2674	0.2918	0.5596
	r^2	0.9554	0.9746	0.9980	0.9980	0.9793
Modified Kelsall	R_∞(%)	32.23	51.76	76.90	71.70	44.06
	k_f (min^{-1})	1.0584	1.2056	1.8889	1.4557	0.7786
	k_s (min^{-1})	2.2204×10^{-14}	1	1	1	0.7786
	φ	0.2692	1.4156×10^{-10}	0.6088	2.609×10^{-10}	0.1114
	r^2	0.9554	0.9746	0.9981	0.9961	0.9790
Fully Mixed	R_∞(%)	28.04	60.32	87.29	80.17	50.76
	k (min^{-1})	0.8603	0.70121	0.60106	0.4612	1
	r^2	0.9185	0.9431	0.9906	0.9870	0.9575

A comparison of different kinetic models fitted to the experimental data for different particle size-classes of raw coal (RC) and waste coal (WC) is shown in Figures 5–14.

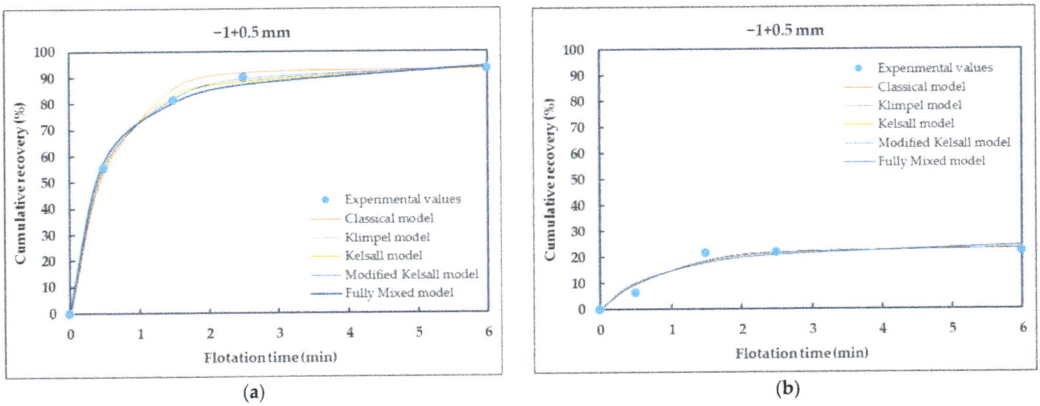

Figure 5. Comparison of different kinetic models fitted to the experimental data of particle size-class −1 + 0.5 mm of (**a**) raw coal (RC) and (**b**) waste coal (WC) samples.

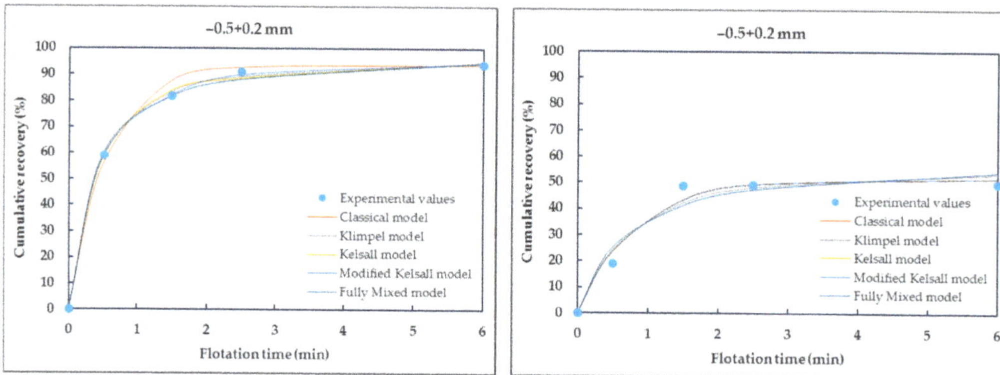

(**a**) (**b**)

Figure 6. Comparison of different kinetic models fitted to the experimental data of particle size-class −0.5 + 0.2 mm of (**a**) raw coal (RC) and (**b**) waste coal (WC) samples.

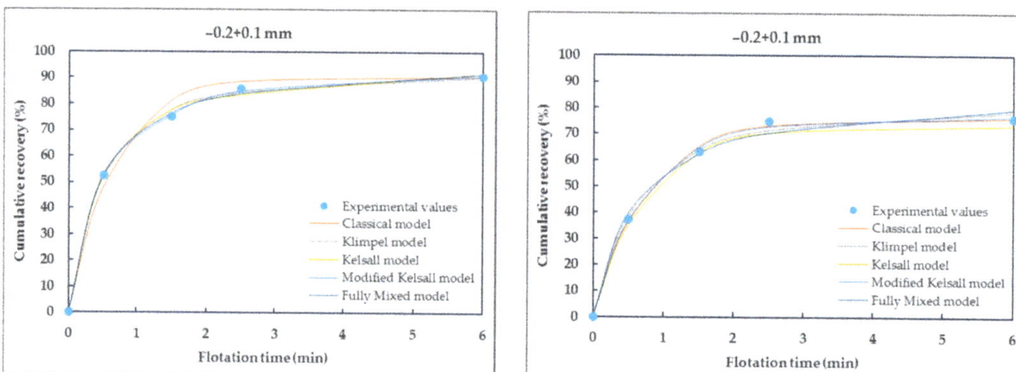

(**a**) (**b**)

Figure 7. Comparison of different kinetic models fitted to the experimental data of particle size-class −0.2 + 0.1 mm of (**a**) raw coal (RC) and (**b**) waste coal (WC) samples.

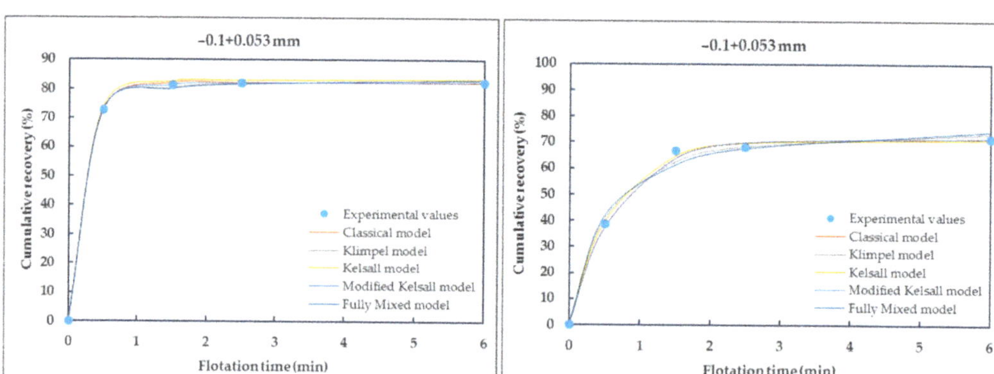

(**a**) (**b**)

Figure 8. Comparison of different kinetic models fitted to the experimental data of particle size-class −0.1 + 0.053 mm of (**a**) raw coal (RC) and (**b**) waste coal (WC) samples.

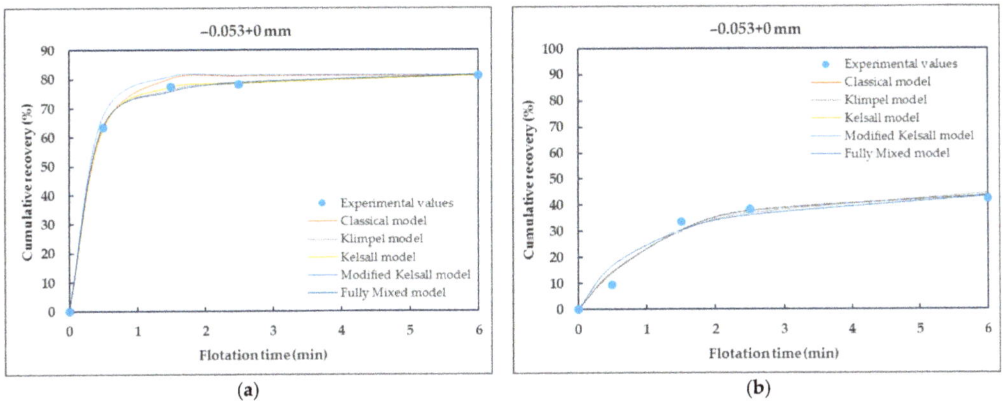

Figure 9. Comparison of different kinetic models fitted to the experimental data of particle size-class −0.053 + 0 mm of (**a**) raw coal (RC) and (**b**) waste coal (WC) samples.

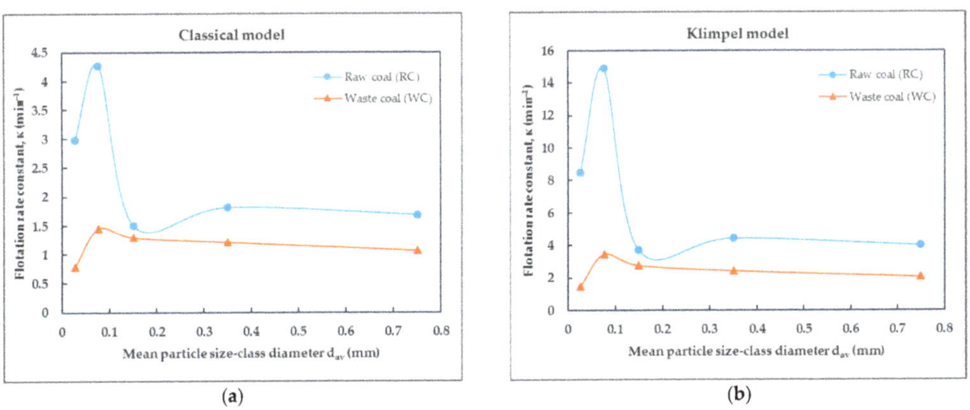

Figure 10. Dependence of the flotation rate constant on particle size of raw (RC) and waste (WC) coal by (**a**) the Classical model and (**b**) the Klimpel model.

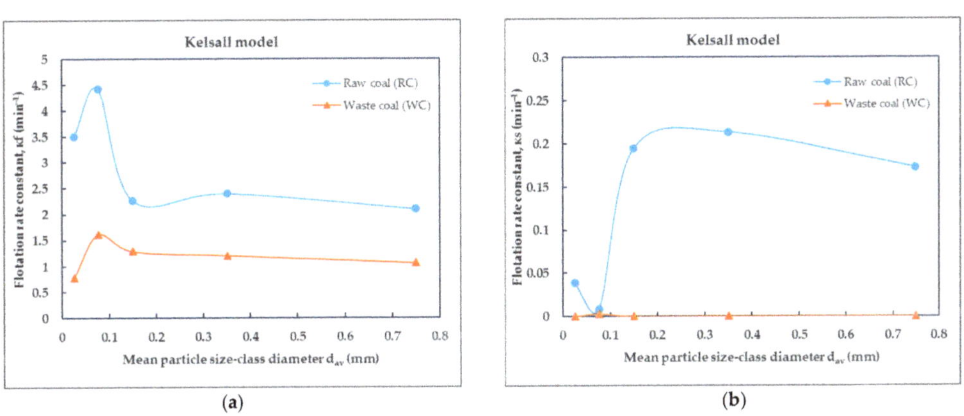

Figure 11. Dependence of the flotation rate constant on particle size of raw (RC) and waste (WC) coal by the Kelsall model: (**a**) the fast and (**b**) the slow flotation rate constant.

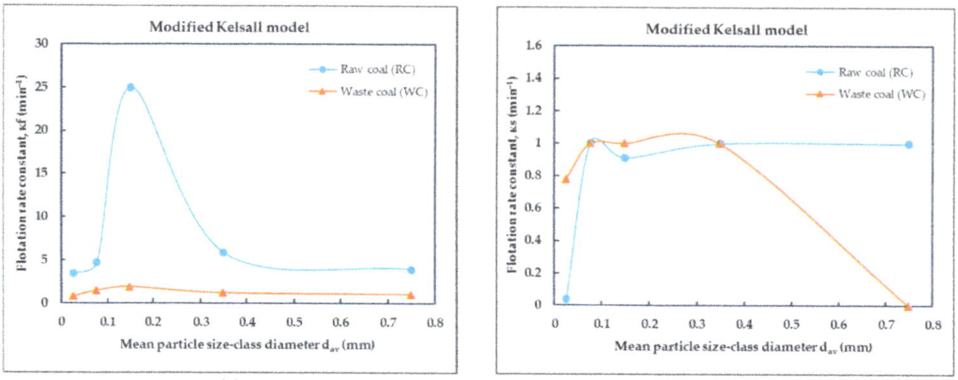

(a) (b)

Figure 12. Dependence of the flotation rate constant on particle size of raw (RC) and waste (WC) coal by the modified Kelsall model: (**a**) the fast and (**b**) the slow flotation rate constant.

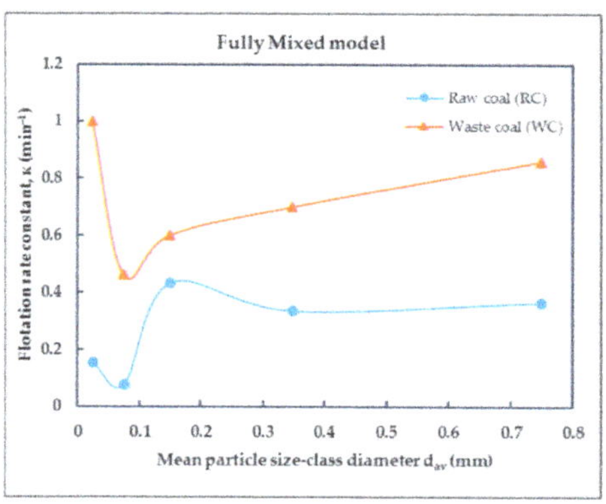

Figure 13. Dependence of the flotation constant on particle size of raw (RC) and waste (WC) coal by the fully mixed model.

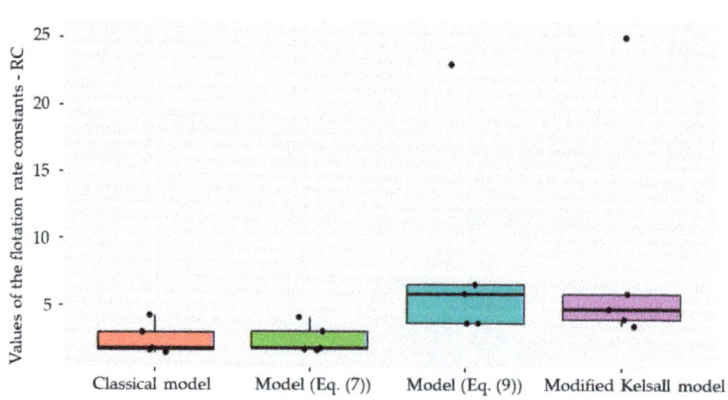

Figure 14. Boxplot for the distribution of actual and predicted values of the flotation kinetics constants by the Classical model for raw and waste coal.

From the presented Tables 5 and 6, it can be seen that all selected flotation kinetics models for describing first-order flotation kinetics show a good correlation with the experimental data. The obtained flotation kinetics curves are similar in shape, with the coefficients of determination being approximately equal to 1. Using the MATLAB tool, the coefficients of determination range from 0.9724 (the Klimpel model) to 1 (the modified Kelsall model) for raw coal and from 0.8609 (the Klimpel model) to 0.9981 (the modified Kelsall model) for waste coal. The poor fit of the Klimpel model (e.g., r^2 = 0.8609 for waste coal) and deviations from the model assumptions most likely arise from the heterogeneity of the waste coal samples, which can lead to variability in flotation behavior and adversely affect the model's accuracy.

Based on the coefficients of determination and the shape of the curves, it is observed that the classical and the modified Kelsall models best describe the first-order flotation kinetics. These models showed better applicability to the experimental data for raw and waste coal, with the coefficients of determination being approximately equal to 1.

Furthermore, the flotation kinetics of raw coal (RC) and waste coal (WC) across different particle size-classes demonstrate significant variation in cumulative recovery and model performance. Coarse particles (+0.5 mm) exhibit rapid flotation kinetics for RC, with most models providing a good fit; however, WC in the same particle size-class shows extremely poor recovery and model mismatch, indicating difficulty in floating coarse waste coal, likely due to poor hydrophobicity.

As particle size decreases, particularly in the (−0.2 + 0.1) mm range, RC samples achieve near-complete recovery with excellent agreement across all kinetic models, highlighting this size-class as the most favorable for flotation efficiency. Conversely, WC continues to exhibit lower recoveries, though performance improves slightly in this intermediate size range. For fine particles (−0.053 + 0 mm), both raw (RC) and waste (WC) coal show reduced flotation kinetics, likely due to increased oxidation and slime effects, with WC being particularly affected.

3.4. *Effect of Coal Particle Size on Flotation Rate Constants*

Based on the values of the flotation rate constants presented in Table 5 for raw coal and Table 6 for waste coal, obtained using different flotation kinetics models and for various particle size-classes, the effect of particle size—expressed by mean particle diameter—on the flotation rate constant of raw coal (RC) and waste coal (WC) is illustrated in the corresponding Figures 10–13. The smooth lines in Figures 10–13 represent trend lines, which connect the points based on the flotation rate constants data. This clarification ensures that the relationship between the variables is accurately represented, whether derived from experimental results or modeled trends.

The flotation kinetics of raw coal (RC) and waste coal (WC) across various particle size-classes reveal significant differences in flotation rate constants and model performance. By comparing the flotation rate constants of different particle size-classes, it can be concluded that all flotation rate constant values for raw coal are higher compared to those for waste coal.

The reduced flotation rate constants observed for waste coal are attributed to oxidation-induced hydrophilicity. Oxidation modifies the coal's surface chemistry by increasing the concentration of oxygen-containing functional groups such as carboxyl, hydroxyl, and carbonyl. These groups enhance the coal's surface hydrophilicity, diminishing its affinity for collector reagents and air bubbles. Consequently, both the flotation efficiency and the flotation rate constants are adversely affected.

Across all models, the Classical and the modified Kelsall models consistently provide better fitting with experimental data for both raw (RC) and waste (WC) coal. Analysis of flotation rate constants reveals that RC consistently peaks around 0.1 mm particle diameter across all models, confirming this as the optimal size for flotation. For WC, however, flotation rate constants remain lower, and there is no clear peak, indicating a fundamentally different and less efficient flotation response.

As can be seen from Figures 10, 11a and 12, the flotation rate constant (k), for raw coal, first increases sharply and then decreases, indicating an optimum particle size (about 0.1 mm), while for waste coal, the flotation rate constant values are lower and increase slightly with increasing particle size. The maximum flotation rate constant (k) values for raw and waste coal are 4.2624 min^{-1} and 1.4557 min^{-1}, respectively, for the particle size-class of (−0.1 + 0.053) mm. The opposite findings were obtained for the Kelsall and fully mixed models. The lowest flotation rate constants were observed in this size-class.

The modified Kelsall model, which differentiates between fast and slow flotation components, offers the most comprehensive fit, capturing the heterogeneous flotation kinetics of both RC and WC. The fast rate constant shows a pronounced maximum for RC near 0.2 mm, followed by a decline with increasing or decreasing size. The slow rate constant shows modest variability in the WC. By applying the modified Kelsall model, the maximum values of the flotation rate constant k for raw and waste coal are 25 min^{-1} and 1.8889 min^{-1} and are obtained for the size-class (−0.2 + 0.1) mm. The fully mixed model also offers valuable insights, particularly for WC, where it more accurately reflects the slower, diffusion-limited nature of flotation.

Depending on the applied flotation kinetics model, it is possible to define the optimal particle size-class in the flotation process of raw and waste coal. For the classic model, the maximum values of the flotation rate constant of raw and waste coa006C are obtained for the size-class (−0.1 + 0.053) mm, while using the modified Kelsall model, it can be observed that the optimal particle size moves towards the size-class (−0.2 + 0.1) mm.

These findings showed that all flotation kinetics models showed good correlation; however, the Classical and the modified Kelsall models provided the best fit.

3.5. The Relationship Between Flotation Kinetics Constant and Coal Particle Size

The relationship between the flotation rate constant and particle size was investigated for both the classical and the modified Kelsall models. According to the results obtained for flotation kinetics parameters by the classical model, the relationship between flotation kinetics constant (k) and average particle size value (d_{sr}) was estimated by Equations (7) and (8):

$$k_{rc} = \frac{a + b\, d_{sr} + c\, {d_{sr}}^2}{1 + d\, d_{sr} + e\, {d_{sr}}^2} \tag{7}$$

where a = 2.728965; b = −41.135466; c = 192.135603; d = −18.675380; e = 102.500493.

$$k_{wc} = a\, d_{sr} + \frac{b}{d_{sr}} + \frac{c}{{d_{sr}}^2} + d \tag{8}$$

where a = −0.1258396; b = 0.0451993; c = −0.0014227; d = 1.1031061.

The relationship between the flotation rate constants and average particle size value for the Modified Kelsall model is expressed only for the fast flotation kinetic (k_f) for both coals through Equations (9) and (10).

$$k_{frc} = a + \frac{d}{c\, d_{sr}\sqrt{2\pi}} \exp\left(-\frac{\left(ln\left(\frac{d_{sr}}{b}\right)\right)^2}{2\,c^2}\right) \tag{9}$$

where a = 3.68734; b = 0.19295; c = 0.37274; d = 3.40367.

$$k_{fwc} = a + b\, d_{sr} + c\, {d_{sr}}^2 + d\, {d_{sr}}^3 + e\, {d_{sr}}^4 + f\, {d_{sr}}^5 \tag{10}$$

where a = 0.03926; b = 29.22031; c = −168.40056; d = 399.43937; e = -426.94274; f = −170.21096.

Analysis of variance of flotation kinetics constants is summarized in Table 7.

Table 7. The values of the flotation kinetics constant of raw (RC) and waste (WC) coal calculated for different models.

	Average Particle Size Value Mean Particle Size-Classes Diameter d_{sr} (mm)	Actual Values by Classical Model	Predicted Values by Model (Equation (7))	Actual Values by Modified Kelsall Model	Predicted Values by Model (Equation (9))
Raw coal (RC)	0.7500	1.6848	1.7907	3.9956	3.6937
	0.3500	1.8084	1.6906	5.9000	6.5930
	0.1500	1.5017	1.7461	25.0000	23.0239
	0.0765	4.2624	4.1271	4.7705	5.8761
	0.0265	2.9839	3.0737	3.5063	3.6874
	Average Particle Size Value d_{sr} (mm)	**Actual Values by Classical Model**	**Predicted Values by Model (Equation (8))**	**Actual Values by Modified Kelsall Model**	**Predicted Values by Model (Equation (10))**
Waste coal (WC)	0.7500	1.0584	1.0665	1.0584	1.0472
	0.3500	1.2056	1.1766	1.2056	1.2504
	0.1500	1.2877	1.3223	1.8889	1.7782
	0.0765	1.4557	1.4412	1.4557	1.4537
	0.0265	0.7786	0.7794	0.7786	0.7026

Boxplots for the distribution of actual and predicted values of the flotation kinetics constants for raw and waste coal are shown in Figures 14 and 15.

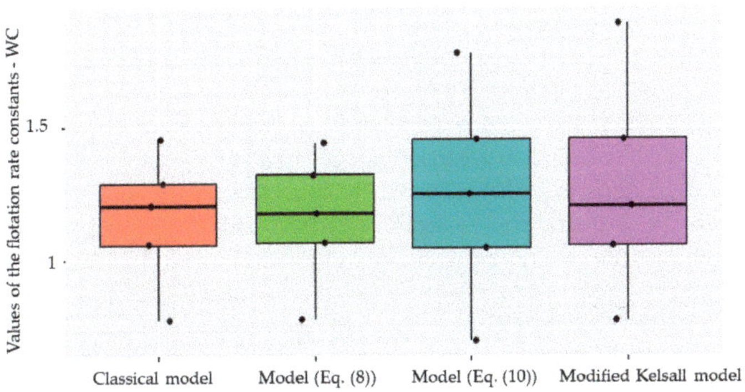

Figure 15. Boxplot for the distribution of actual and predicted values of the flotation kinetics constants by the Modified Kelsall model for raw and waste coal.

For statistical data analysis, the Kruskal–Wallis (K-W) test and the Shapiro–Wilk (S-W) test were employed. The results of the statistical analysis are presented in Table 8.

Table 8. The statistical test parameters for the raw and waste coal samples.

	Statistical Test Parameters	Actual Values by Classical Model	Predicted Values by Model (Equation (7))	Actual Values by Modified Kelsall Model	Predicted Values by Model (Equation (9))
Raw coal (RC)	Mean value	2.44824	2.48565	8.63538	8.57482
	Standard deviation	1.169004	1.084229	9.192669	8.180752
	Coefficient of variation	47.75%	43.62%	106.45%	95.40%
	Shapiro–Wilk normality test	+	+	-	-
	95% confidence interval	(0.9967; 3.8998)	(1.1394; 3.8319)		
	Statistical Test Parameters	**Actual Values by the Classical Model**	**Predicted Values by Model (Equation (8))**	**Actual Values by Modified Kelsall Model**	**Predicted Values by Model (Equation (10))**
Waste coal (WC)	Mean value	1.1572	1.1572	1.2774	1.24642
	Standard deviation	0.2557	0.2546	0.4206	0.4066
	Coefficient of variation	22%	22%	32.9%	32.6%
	Shapiro–Wilk normality test	+	+	+	+
	95% confidence interval	(0.8397; 1.4747)	(0.8411; 1.4733)	(0.7552; 1.7997)	(0.7416; 1.7512)

Based on the Shapiro–Wilk normality test (see the row in Table 8—"Shapiro–Wilk normality test = '+' or '-'"), the four samples, obtained by four different methods, were checked for normality. For the sample that failed the normality test, we concluded that a parametric test like ANOVA could not be applied. Instead, we used the nonparametric alternative, the Kruskal–Wallis test. The results of the Kruskal–Wallis test are as follows: Kruskal–Wallis chi-squared = 10.109, df = 3, *p*-value = 0.01767. From this, we can conclude that there are significant differences between the models, as the distribution of the models is not the same.

Since all four samples (obtained by four methods) have normal distribution (as indicated by the Shapiro–Wilk normality test in Table 8—"Shapiro–Wilk normality test = '+'"), we applied the ANOVA test. The calculated F-value was 0.162, which is less than the critical value F_(3;16;0.05). Therefore, the ANOVA test indicates that the means of all four models are equal, suggesting no significant difference between them.

Comparison between actual and predicted values of the flotation kinetics constants for raw and waste coal, listed in Table 7, is given in Figures 16 and 17.

The coefficient of determination (r^2) and coefficient of correlation (ρ) were used for statistical data analysis. The statistical parameters for raw (RC) and waste (WC) coal are given in Tables 9 and 10.

Table 9. Correlation between certain models for RC.

Flotation Kinetics Model	Statistical Parameters	Model (Equation (7))	Model (Equation (9))
Classical	Coefficient of determination (r^2)	0.9852	
	Coefficient of correlation (ϱ)	0.9926	
Modified Kelsall	Coefficient of determination (r^2)		0.9946
	Coefficient of correlation (ϱ)		0.9973

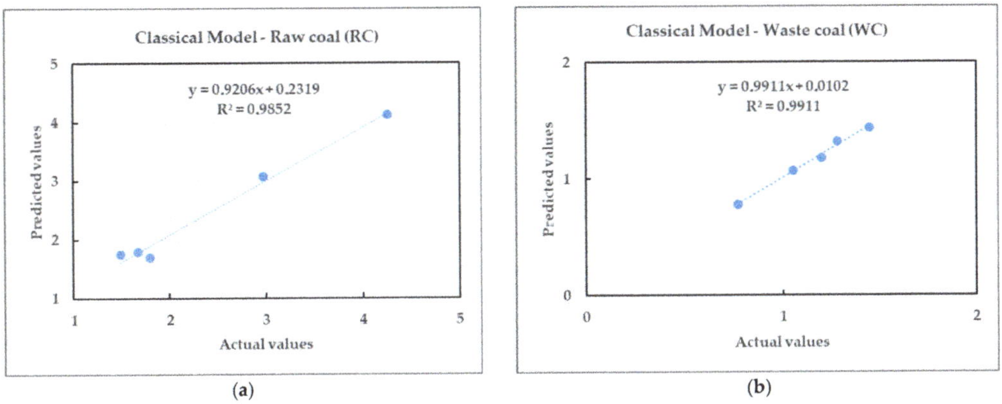

Figure 16. Relationship between the actual and predicted values of the flotation rate constants for the classical model.

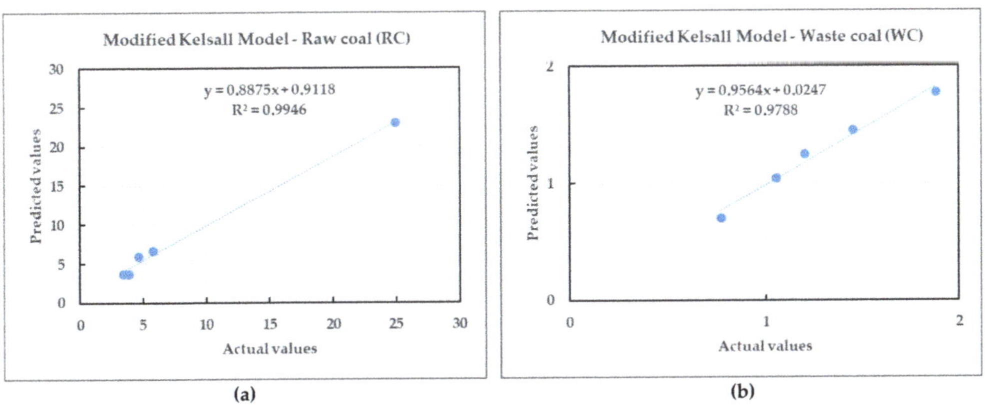

Figure 17. Relationship between the actual and predicted values of the fast flotation rate constants for the modified Kelsall model.

Table 10. Correlation between certain models for WC.

Flotation Kinetics Model	Statistical Parameters	Model (Equation (8))	Model (Equation (10))
Classical	Coefficient of determination (r^2)	0.9911	
	Coefficient of correlation (ϱ)	0.9955	
Modified Kelsall	Coefficient of determination (r^2)		0.9788
	Coefficient of correlation (ϱ)		0.9893

The coefficient of determination (r^2) values for the flotation kinetics constant for raw and waste coal by the classical model were 0.9852 and 0.9911, respectively. Based on the observed trend, the proposed mathematical functions given in Equations (7) and (8) are a good fit for both coals. The coefficient of correlation (ρ) values for raw and waste coal were 0.9926 and 0.9955, respectively. The results indicate that both models are capable of predicting the flotation kinetics constant with minimal errors and high accuracy.

The r^2 values for the fast flotation kinetics constant for raw and waste coal by the modified Kelsall model were 0.9946 and 0.9788, respectively. Based on the observed trend, both mathematical functions given in Equations (9) and (10) are a good fit for both coals. The analysis shows the coefficient of correlation (ρ values obtained for raw and waste coal were 0.9973 and 0.9893, respectively). r^2 and ρ values showed that both mathematical functions give almost better estimates with smaller errors.

It was observed that the flotation kinetics constant (k) for individual particle sizes of the raw and waste coal could be predicted satisfactorily. Overall, these findings confirmed the importance of investigating the relationship between particle size and flotation rate constants across different kinetic models in order to optimize flotation performance, particularly in coal flotation.

4. Conclusions

The results of the study presented in this paper show that the classical and the modified Kelsall models can be successfully used as an effective tool for modeling the complex relation between the particle size and flotation rate constants in the flotation of raw and waste coal. The coefficient of determination (r^2) values were obtained from 0.9904 to 0.9998 for raw coal and from 0.9555 to 0.9980 for waste coal, indicating a good relation.

The correlation between the flotation rate constant (k) and the average particle size value (d_{sr}) depends on the type of coal. For the classical model, the maximum flotation rate constant (k) values for raw and waste coal are 4.2624 min^{-1} and 1.4557 min^{-1}, respectively, for the particle size-class of (−0.1 + 0.053) mm. By the modified Kelsall model, the maximum values of the fast flotation rate constant k for raw and waste coal are 25 min^{-1} and 1.8889 min^{-1} and are obtained for the particle size-class (−0.2 + 0.1) mm.

This study shows that particle size has a key impact on the flotation performance of waste coal, with the (–0.2 + 0.1) mm size-class identified as optimal for both flotation rate and overall recovery. Recognizing this optimal size-class has significant industrial value, as it provides clear guidelines for more efficient coal classification in preparation plants, enabling improvements in flotation process efficiency, reductions in reagent consumption, and enhancements in concentrate quality. Additionally, applying this approach contributes to more sustainable and cost-effective utilization of oxidized coals.

Considering the established significance of particle size for flotation kinetics, the proposed mathematical models given in Equations (7) and (8) (the classical model) and Equations (9) and (10) (the modified Kelsall model) provide a precise mathematical relationship between the flotation kinetics constant (k) and average particle size value (d_{sr}). Statistical analysis shows that these models can support the prediction and optimization of flotation kinetics with a high degree of accuracy in practical conditions.

Author Contributions: Conceptualization, J.S., I.Đ. and D.T.; methodology, J.S., I.Đ. and D.T.; software, I.Đ. and D.T.; validation, J.S. and Z.Š.; formal analysis, J.S., I.Đ. and D.T.; investigation, J.S. and I.I.; resources J.S. and I.I.; writing—original draft preparation, J.S., I.Đ., D.T. and Z.Š.; writing—review and editing, J.S., I.Đ., D.T. and Z.Š.; visualization, J.S. and Z.Š.; supervision, J.S. All authors have read and agreed to the published version of the manuscript.

Funding: This research was funded by the Ministry of Science, Technological Development and Innovation of the Republic of Serbia, within the funding of the scientific research work at the University of Belgrade, Technical Faculty in Bor, according to the contract with registration number 451-03-137/2025-03/200131.

Data Availability Statement: All data are available upon request.

Acknowledgments: The research presented in this paper was performed with the financial support of the Ministry of Science, Technological Development and Innovation of the Republic of Serbia, within the funding of the scientific research work at the University of Belgrade, Technical Faculty in Bor, according to the contract with registration number 451-03-137/2025-03/200131.

Conflicts of Interest: The authors declare that they have no known competing financial interests or personal relationships that could have appeared to influence the work reported in this paper.

References

1. Polat, M.; Polat, H.; Chander, S. Physical and chemical interactions in coal flotation. *Int. J. Miner. Process.* **2003**, *72*, 199–213. [CrossRef]
2. Brozek, M.; Mlynarczykowska, A. An analysis of effect of particle size on batch flotation of coal. *Physicochem. Probl. Miner. Process.* **2013**, *49*, 341–356.
3. LaskowskI, J.S. *Coal Flotation and Fine Coal Utilization*; Elsevier: Amsterdam, The Netherlands, 2001.
4. Polat, M.; Arnold, B.; Chander, S.; Hogg, R.; Zhou, R. Coal flotation kinetics-effect of particle size and specific gravity. *Coal Sci. Technol.* **1993**, *24*, 161–170.
5. Lynch, A.J.; Johnson, N.W.; Manlapig, E.V.; Thorne, C.G. *Mineral and Coal Flotation Circuits: Their Simulation and Control*; Elsevier: Amsterdam, The Netherland, 1981.
6. Polat, M.; Chander, S. First-order flotation kinetics models and methods for estimation of the true distribution of flotation rate constants. *Int. J. Miner. Process.* **2000**, *58*, 145–166. [CrossRef]
7. Drzymala, J. *Mineral Processing. Foundations of Theory and Practice of Minerallurgy*, 2nd ed.; Ofic. Wyd. PWr: Wroclaw, Poland, 2018.
8. Sokolović, J.; Miskovic, S. The effect of particle size on coal flotation kinetics: A review. *Physicochem. Probl. Miner. Process.* **2018**, *54*, 1172–1190.
9. Panopoulos, G.; King, R.P.; Juckes, A. The effect of particle-size distribution on the flotation of two South African coals. *J. S. Afr. Inst. Min. Metall.* **1986**, *86*, 141–152.
10. Vanangamudi, M.; Rao, T.C. Modelling of batch coal flotation operation. *Int. J. Miner. Process.* **1986**, *16*, 231–243. [CrossRef]
11. Vanangamudi, M.; Kumar, S.S.; Rao, T.C. Effect of fines content on the froth flotation of coal. *Powder Technol.* **1989**, *58*, 99–105. [CrossRef]
12. Rao, T.C.; Govindarajan, B.; Vanangamudi, M. A kinetic model for batch coal flotation. *Miner. Eng.* **1989**, *2*, 403–414. [CrossRef]
13. Mohns, C.A. *Effect of Particle Size on Coal Flotation Kinetics*; Queen's University: Kingston, ON, Canada, 1998.
14. Chaves, A.P.; Ruiz, A.S. Considerations on the kinetics of froth flotation of ultrafine coal contained in tailings. *Int. J. Coal Prep. Util.* **2009**, *29*, 289–297. [CrossRef]
15. Abkhoshk, E.; Kor, M.; Rezai, B. A study on the effect of particle size on coal flotation kinetics using fuzzy logic. *Expert Syst. Appl.* **2010**, *37*, 5201–5207. [CrossRef]
16. Kor, M.; Abkhoshk, E.; Gharibie, K.; Shafaei, S.Z. An investigation of the particle size effect on coal flotation kinetics using multivariable regression. *J. Min. Environ. Issues* **2010**, *1*, 41–47.
17. Li, Y.; Zhao, W.; Gui, X.; Zhang, X. Flotation kinetics and separation selectivity of coal size fractions. *Physicochem. Probl. Miner. Process.* **2013**, *49*, 387–395.
18. Xia, W.; Xie, G.; Liang, C.; Yang, J. Flotation behavior of different size fractions of fresh and oxidized coals. *Powder Technol.* **2014**, *267*, 80–85. [CrossRef]
19. Ni, C.; Xie, G.; Jin, M.; Peng, Y.; Xia, W. The difference in flotation kinetics of various size fractions of bituminous coal between rougher and cleaner flotation processes. *Powder Technol.* **2016**, *292*, 210–216. [CrossRef]
20. Bu, X.; Xie, G.; Chen, Y.; Ni, C. The order of kinetic models in coal fines flotation. *Int. J. Coal Prep. Util.* **2017**, *37*, 113–123. [CrossRef]
21. Xia, W.C.; Yang, J.G. Experiment design of oily bubble in oxidized coal flotation. *Gospod. Surowcami Miner.* **2013**, *29*, 129–135.
22. Bu, X.; Wang, X.; Zhou, S.; Li, B.; Zhan, H.; Xie, G. Discrimination of six flotation kinetic models used In the conventional flotation and carrier flotation of −74 μm coal fines. *Acs. Omega.* **2020**, *5*, 13813–13821. [CrossRef] [PubMed]
23. Shahbazi, B.; Chelgani, S.C. Modeling of fine coal flotation separation based on particle characteristics and hydrodynamic conditions. *Int. J. Coal Sci. Technol.* **2016**, *3*, 429–439. [CrossRef]
24. Norori-McCormac, A.; Brito-Parada, P.R.; Hadler, K.; Cole, K.; Cilliers, J.J. The effect of particle size distribution on froth stability in flotation. *Sep. Purif. Technol.* **2017**, *184*, 240–247. [CrossRef]
25. Liao, Y.; Cao, Y.; Liu, C.; Zhao, Y.; Zhu, G. Comparison of the effect of particle size on the flotation kinetics of a low-rank coal using air bubbles and oily bubbles. *J. S. Afr. Inst. Min. Metall.* **2017**, *117*, 561–566. [CrossRef]
26. Han, H.; Liu, A.; Wang, C.; Yang, R.; Li, S.; Wang, H. Flotation kinetics performance of different coal size fractions with nanobubbles. *Int. J. Miner. Metall. Mater.* **2022**, *29*, 1502–1510. [CrossRef]
27. Sahoo, S.K.; Suresh, N.; Varma, A.K. Determining the best particle size-class for flotation of a high ash coal. *Int. J. Coal Prep. Util.* **2020**, *40*, 755–765. [CrossRef]
28. Bahrami, A.; Ghorbani, Y.; Gulcan, E.; Kazemi, F.; Kakaei, H.; Farajzadeh, S. Effects of particle size distribution on flotation kinetics of bituminous coal. *Iran. J. Chem. Eng. (IJChE)* **2020**, *17*, 3–13.

29. Şahbaz, O.; Demir, M.K. Effects of frothers and particle size on the flotation kinetics of the Jameson Cell. *Physicochem. Probl. Miner. Process.* **2020**, *56*, 829–838. [CrossRef]
30. Karaca, S.; Ucar, A.; Sahbaz, O. Kinetic modelling of flotation column and Jameson cell in coal. *Physicochem. Probl. Miner. Process.* **2022**, *58*, 152848. [CrossRef]
31. Lu, T.; Deng, Z.; Tang, Y.; Cheng, W. Effects of particle size on the flotation behavior of coal slime. *Energy Explor. Exploit.* **2023**, *41*, 1344–1355. [CrossRef]
32. Kazemi, F.; Bahrami, A.; Ghorbani, Y.; Danesh, A.; Abdollahi, M.; Falah, H.; Salehi, M. The interaction and synergic effect of particle size on flotation efficiency: A comparison study of recovery by size, and by liberation between lab and industrial scale data. *Rud.-Geološko-Naft. Zb.* **2023**, *38*, 1–12. [CrossRef]
33. Chang, Z.; Chen, X.; Peng, Y. Understanding and improving the flotation of coals with different degrees of surface oxidation. *Powder Technol.* **2017**, *321*, 190–196. [CrossRef]
34. Xia, W.; Yang, J.; Liang, C. A short review of improvement in flotation of low rank/oxidized coals by pretreatments. *Powder Technol.* **2013**, *237*, 1–8. [CrossRef]
35. Wen, B.; Xia, W.; Sokolovic, J.M. Recent advances in effective collectors for enhancing the flotation of low rank/oxidized coals. *Powder Technol.* **2017**, *319*, 1–11. [CrossRef]
36. Sokolović, J.; Stanojlović, R.; Marković, Z. The effects of pretreatment on the flotation kinetics of waste coal. *Int. J. Coal Prep. Util.* **2012**, *32*, 130–142. [CrossRef]
37. Sokolovic, J.; Stanojlovic, R.; Markovic, Z. Activation of oxidized surface of anthracite waste coal by attrition. *Physicochem. Probl. Miner. Process.* **2012**, *48*, 5–18.
38. *ASTM D3174-04(2010)*; Standard Test Method for Ash in the Analysis Sample of Coal and Coke from Coal. ASTM International: West Conshohocken, PA, USA, 2010.
39. *ISO 589:2008*; Hard Coal—Determination of Total Moisture. International Organization for Standardization: Geneva, Switzerland, 2008.
40. *ASTM D3175-11*; Standard Test Method for Volatile Matter in the Analysis Sample of Coal and Coke. ASTM International: West Conshohocken, PA, USA, 2011.
41. *ASTM D5373-08*; Standard Test Methods for Instrumental Determination of Carbon, Hydrogen, and Nitrogen in Laboratory Samples of Coal. ASTM International: West Conshohocken, PA, USA, 2008.
42. *ASTM D4239-11*; Standard Test Methods for Sulfur in the Analysis Sample of Coal and Coke Using High-Temperature Tube Furnace Combustion Methods. ASTM International: West Conshohocken, PA, USA, 2011.
43. *ASTM D5865-11*; Standard Test Method for Gross Calorific Value of Coal and Coke. ASTM International: West Conshohocken, PA, USA, 2011.
44. Zuniga, H.G. Flotation recovery is an exponential function of its rate. *Boln. Soc. Nac. Min. Santiago Chile* **1935**, *47*, 83–86.
45. Klimpel, R.R. *Selection of Chemical Reagents for Flotation. Mineral Processing Plant Design*, 2nd ed.; SME-AIME: New York, NY, USA, 1980; pp. 907–934.
46. Kelsall, D.F. Application of probability assessment of flotation systems. *Trans. Am. Soc. Min. Metall. Eng.* **1961**, *70*, 191–204.
47. Jowett, A. Resolution of flotation recovery curves by a differential plot method. *Trans. Am. Soc. Min. Metall. Eng.* **1974**, *85*, C263–C266.
48. Imaizumi, T.; Inoue, T. Kinetic consideration of froth flotation. In Proceedings of the 6th International Mineral Processing Congres, Cannes, France, 29 May–3 June 1965; pp. 563–579.
49. Brezani, I.; Zelenak, F. MatLab Tool for Modeling First Order Flotation. 2010. Available online: https://www.mathworks.com/matlabcentral/fileexchange/28703-flotation-kinetics-modeling (accessed on 31 August 2011).

minerals

MDPI

Article

Technological Properties Contrast of Galena, Sphalerite, Carbonaceous Material and Choice of Flotation Technology

Akim Yergeshev [1,2], Rustam Tokpayev [1,*], Marina Karmeeva [2], Tamina Khavaza [1,*], Nazymarzu Yergesheva [2], Azhar Atchabarova [1], Mikhail Nauryzbayev [1] and Vladislava Ignatkina [2]

1 Center of Physico-Chemical Methods of Research and Analysis, Al-Farabi Kazakh National University, Almaty 050000, Kazakhstan; akim9797@mail.ru (A.Y.); azhar.atchabarova@mail.ru (A.A.); nauryzbayev@cfhma.kz (M.N.)

2 Department of Mineral Processing and Industrial Wastes Recycling, College of Mining, National University of Science and Technology "MISIS" (NUST MISIS), Moscow 119049, Russia; m2001044@edu.misis.ru (M.K.); nazymarzu.zharolla@mail.ru (N.Y.); woda@mail.ru (V.I.)

* Correspondence: rustamtokpaev@mail.ru (R.T.); k.tamina@mail.ru (T.K.); Tel.: +7-707-522-44-88 (R.T.); +7-747-280-93-98 (T.K.)

Abstract

The presence of galena, sphalerite (cleiophane), and Carbonaceous Material (CM) in sulphide ore complicates the application of a direct-differential flotation flowsheet due to increased mutual interactions between both marketable concentrates and final tailings. Flotation tests, measurements of electrokinetic (zeta) potential, adsorption of sulphydric collectors, and colorimetric indicators were employed to elucidate the cause-and-effect relationships underlying the reduction in contrast of the flotation properties of galena and cleiophane surfaces. It was established that galena and cleiophane exhibit comparable flotation responses when using diesel oil within a pH range of 6–8. While high galena recovery is anticipated, the similar recovery of cleiophane is attributed to the ZnS zeta potential approaching zero in this pH interval. Experimental results demonstrated a distinct difference in the flotation behavior of galena and cleiophane, both with natural surface oxidation and following the removal of sulphoxy films. The application of Carbonaceous Material depressants derived from wood processing by-products (lignin-sulphonates) resulted in a significant decrease in sphalerite recovery. Although the flotation rate constant for Carbonaceous Material in the presence of lignin-sulphonate-based depressants decreases, the overall recovery to concentrate increases over time. The implementation of a bulk-differential flowsheet, involving the preliminary removal of CM prior to the bulk Pb-Zn flotation of lead-zinc sulphide ore, has been demonstrated to be effective.

Keywords: galena; sphalerite; carbonaceous material; flotation; adsorption; ZetaPotential; depression; sulphydric collectors; velocity constant; technological scheme

1. Introduction

Sulphides in most sulphide ores are disseminated; thus, flotation is the principal process for the recovery of sulphides from ores into saleable concentrates.

Sphalerite (ZnS) is the primary industrial zinc mineral. The composition of impurities significantly influences its technological properties and is reflected in the mineral's color, ranging from light yellow in the purest ZnS variety (cleiophane) to dark brown in marmatite, where the iron content varies from 5 to 22%. It is well established that the adsorption and flotation activity of sphalerite increase in the presence of copper as an impurity or after copper sulfate treatment in an alkaline environment. Other heavy-metal ions may also activate sphalerite flotation [1–4].

Galena (PbS) is the primary industrial lead mineral and is the most extensively studied among sulphidic minerals, owing to the availability of monomineral nodules. Galena is a refractory sulphide, which results in more stable surface properties during flotation studies [5–8]. In general, the processing of lead ores containing granular or unevenly disseminated galena employs combined gravity-flotation flowsheets.

Glembotsky and Dmitrieva [9] investigated 28 galena samples containing between 86.07% and 62.16% lead; however, they found no clear correlation between the type of conductivity and the flotation activity of galena. The authors identified two groups of galena. The first group, characterized by a minimal amount of impurities, exhibits active interaction with ethyl xanthogenate and dithiophosphate. The second group typically displays higher surface heterogeneity and a greater tendency toward oxidation, which is accompanied by weak interaction with ethyl xanthogenate and an inability to be recovered by dithiophosphate.

The flotation activity of galena depends on the pH of the environment. When using ethyl xanthogenate (X) and diethyl dithiophosphate (DPT) individually at pH = 9, the recovery is comparable. However, when both collectors are used together, the order of addition influences recovery; higher recovery is achieved when xanthogenate is added first and constitutes a smaller proportion of the mixture (X:DPT = 10:90). Both X and DPT are adsorbed onto galena in ionic form [10].

Researchers have observed changes in galena flotation activity in the presence of pyrite, beginning at the grinding stage, where the intensity of oxidation increases rapidly and flotation properties are altered [11].

In recent decades, in addition to established factors influencing the flotation properties of sulphides—such as the composition of mineral impurities, surface oxidation rate, and the presence of pyrite—the occurrence of Carbonaceous Material (CM) in ores has also been recognized as a significant factor. The presence of Carbonaceous Material, a naturally hydrophobic mixed-phase mineral, leads to substantial mutual losses of both galena and sphalerite in saleable concentrates [12,13].

Direct-differential flotation using cyanide in the reagent scheme to depress all sulphides except galena remains the most widely applied technology for the flotation of lead-zinc ores. Non-cyanide reagent regimes for direct-differential flotation, based on combinations of zinc sulfate and sulphoxy compounds (Na_2S, NaHS, $NaHSO_3$, Na_2SO_3), are not always effective in depressing the flotation activity of low-impurity sphalerite (cleiophane). The bulk-differential flotation flowsheet yields the best technological outcomes. Carbonaceous polymetallic sulphidic ores present significant challenges for beneficiation [14–19].

Polymetallic ore from the Shalkiya deposit (Kazakhstan) contains two types of naturally hydrophobic nodules: CM and galena. In addition, sphalerite occurs as its purest natural variety, cleiophane, which exhibits high flotation ability. The objectives of this research are as follows:

To experimentally determine the causes of the low contrast in the flotation of galena, sphalerite and Carbonaceous Material.

To provide justification for the use of a bulk-differential flotation scheme for the beneficiation of complex sulphidic ore from the Shalkiya deposit (Kazakhstan).

2. Materials and Methods

The study objects include monomineral fractions of galena and sphalerite, as well as CM recovered from polymetallic ore of the Shalkiya lead-zinc deposit.

Pure mineral and Carbonaceous Material (CM). Table 1 presents the properties of sulphides and Carbonaceous Material (CM). The chemical composition of sulphides was determined with X-ray fluorescence spectroscopy (XRF) using a Shimadzu XRF-1800 (Shimadzu, Kyoto, Japan). The specific surface area of each fraction was measured by nitrogen low-temperature adsorption (BET method) with a Nova 2200E (Quantachrome Instruments, Boynton Beach, FL, USA).

Table 1. Chemical composition of galena, sphalerite and Carbonaceous Material (CM).

	Fraction, μm	Specific Surface, g/m^2	Elemental Composition, %										
			Pb	Zn	S	Fe	Cu	Cd	Al	Si	C	Other	Total
PbS	−44 + 10	0.520	83.86	1.17	12.65	0.69	0.13	-	0.04	0.19	-	1.26	100.00
	−71 + 41	abs											
ZnS	−44 + 10	0.864	1.50	63.10	31.07	1.20	0.10	0.40	0.03	0.90	-	1.70	100.00
	−71 + 41	abs											
CM	−71 + 41	12.93	1.39	4.71	6.04	4.56			0.30	5.10	10.60	67.3 *	100.00

Note * Na, Ca, Mg, O.

X-ray diffraction analysis of sulphide samples was performed using a Bruker D2 Phaser X-ray powder diffraction system. The monomineral galena fraction was found to contain 98% PbS by mass (space group Fm-3m), approximately 1% ZnS (space group F-43m) and minor amounts of other phases (~1%). X-ray diffraction analysis of ZnS samples showed 97% cubic ZnS (space group F-43m), minor PbS (space group Fm-3m) and traces of crystalline SiO_2.

Figure 1 presents the IR spectrum in the range of 1900–400 cm^{-1} for CM recovered from lead-zinc ore of the Shalkiya deposit.

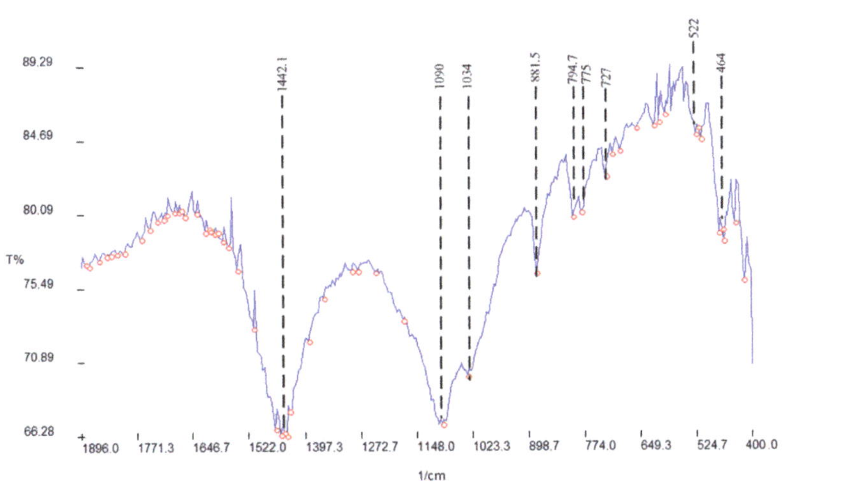

Figure 1. IR Transmittance spectrum of Carbonaceous Material (CM), $v_{mid.}$ ± 1 cm^{-1}, red circles denote the measured data points; the blue—spectrum.

Typical IR absorption bands indicate the presence of carbonates and feldspars associated with rock-forming silicates and quartz. The strong bands with minimal transmission at 1452 and 1440 cm^{-1} correspond to C–O stretching vibrations (ν_{CO}) in carbonates; broad, intense bands with minima at 1090 and 1034 cm^{-1} are attributed to Si–O stretching vibrations; bands at 881.5 cm^{-1}, doublet at 794.7 and 775 cm^{-1} and several bands of varying intensity at 727, 522 and 464 cm^{-1} are related to deformation and out-of-plane vibrations of X–O bonds, where X can be Si, C or other elements present in the carbonaceous sample. The IR spectra of the CM from the ore are consistent in their main characteristic bands with those presented in the study by Lee et al. [20].

Elemental composition of the sample surfaces (Table 2) was determined using X-ray photoelectron spectroscopy (XPS).

Table 2. Elemental composition of Carbonaceous Material (CM) samples surface, ±standard deviation, %.

	C ±1.0	O ±1.0	Si ±0.5	Mg ±0.5	Al ±0.5	Ca ±0.5	P ±0.1	N ±0.3	F ±0.5	Zn ±0.5	Fe ±0.3	K ±0.1	Na ±0.1	S ±0.2	Pb ±0.1
CM	49.2	34.5	5.1	2.5	1.9	2.5	-	-	-	2.8	0.4	-	-	0.7	0.4

Sulphide fractions were prepared from lump samples by manual grinding in an agate mortar and dry classification. The −10 μm fraction was removed by sedimentation. To remove sulphoxy films prior to flotation or adsorption, sulphide aliquots were soaked in 1% NaOH for 24 h, then washed to neutral pH using five volumes of distilled water.

CM was recovered from the Shalkiya deposit ore by flotation in a mechanical cell without collectors or frothers, followed by air drying and classification into narrow fractions.

Polymetallic ore. The mineral composition of polymetallic ore from the Shalkiya deposit includes 12.81% ore minerals, comprising 1.57% galena and 6.94% sphalerite (Zn content in grains not less than 64%). The remaining ore minerals are primarily pyrite. Non-ore minerals constitute 87.19% of the mass, including 43.97% carbonates, 38.17% quartz, and 1.31% carbonaceous substance [21].

Reagents. The reagents used in this research are listed in Tables 3 and 4.

Table 3. Properties of reagents used in the research of PbS, ZnS and CM.

Technological Purpose	Reagent	Empirical Formula	Molar Mass, g/mol	Activity, %
Sulphydryl ionized collectors	Potassium ethyl xanthogenate (KEtX)	$C_2H_5OS_2K$	160.30	93.0
	Potassium butyl xanthogenate (KButX)	$C_5H_9OS_2K$	188.36	91.5
	Sodium diisobuthyl dithiophosphate (DPT)	$C_8H_{18}O_2PS_2Na$	264.30	56.0
Non-ionized collector	O-isopropyl-N ethyltionocarbamate (TC)	$C_6H_{13}NOS$	147.24	92.0
	Diesel oil «winter» (DO)	$C_{12.343}H_{23.889}$	172.3	98.0
Modifiers Environment regulator	Sodium hydroxide lime milk	NaOH $Ca(OH)_2$	40.00 74.09	100.0 100.0
Carbonaceous Material (CM) depressor	lignin-sulphonate	$C_{20}H_{24}O_{10}S_2Na_2$ (variable valent)	534.51	100.0
	P2	Mixture of sodium polymethylenenaphtaline-sulphonate (80%), lignin-sulphonate (10%), etc.	1640	100.0
Frothing agent	Pine oil	$C_{10}H_{16}$	136.23	>96

Table 4. Properties of reagents used in research of PbS-ZnS-CM-S ore sample.

Technological Purpose	Reagent
Sulphydryl ionized collectors	Potassium butyl xanthogenate (KButX) Sodium diisobuthyl dithiophosphate (DPT)
Non-ionized collector	Diesel oil «winter» (DO), O-isopropyl-N ethyltionocarbamate (TC or Z 200)
Modifiers	
Depressors	
Reducing flotation activity of PbS, ZnS	NaHS
Reducing flotation activity of ZnS	$ZnSO_4$ + $NaHSO_3$
Reducing flotation activity of FeS_2	$Ca(OH)_2$
Sphalerite activator	$CuSO_4$
Desorption	NaHS+ active carbon

2.1. Research Methods

2.1.1. Flotation

The research employed several flotation study methods: microflotation in a Hallimond tube, froth flotation of monomineral and ore samples, and kinetic studies through fractionated froth removal.

Recovery (ε%) of minerals or elements was determined using the equation below:

$$\varepsilon = \gamma \times \beta / \alpha \quad (1)$$

γ—yield of flotation product, %

β, α—content, %

In flotation tests using pure mineral samples, the mineral content in both the flotation products (β, %) and the initial feed (α, %) is assumed to be identical ($\beta = \alpha = 100\%$). Therefore, the recovery (ε) is equal to the yield (γ).

Microflotation in a Hallimond tube. Microflotation experiments were conducted in a 50 mL Hallimond tube using 0.5 g of mineral aliquot with particle size fraction −0.41 +0.10 μm and a solid-to-liquid ratio (S:L) of 1:100. The tube design allows flotation with discrete air bubbles generated via a system of glass capillaries in connecting tubes, with air displaced with water from a Drechsel bottle. Flotation duration was controlled by the volume of displaced air. The suspension was stirred using a magnetic stirrer without the addition of frothing agents; reagent conditioning time was set at 15 min.

Froth flotation of monomineral fractions was performed in a mechanical flotation cell. Each test used 5 g aliquots of sulphide or CM in the −71 + 41 μm fraction, with a cell volume of 50 mL. The mineral aliquot was first conditioned with pH-adjusted water for 3 min, followed by sequential addition and mixing of collector and frother for 1 min. In the mechanical cell, suspension mixing and air dispersion were generated by an impeller, and a foam removal device was used to collect the froth. Air flow rate was maintained at 20%, and water level was regulated with pH-adjusted water.

Figure 2 illustrates the dependence of PbS recovery on the flotation method.

As shown in Figure 2, the recovery trends for galena by froth flotation (1.2 method) and Hallimond tube microflotation (1.1 method) are similar in character, although the absolute recovery levels differ (95% vs. 40%). The two methods differ in terms of air dispersion conditions, the presence or absence of a frother, and the particle size fractions used. Both methods reach a plateau at a potassium butyl xanthogenate (KButX) concentration of 10^{-4} M.

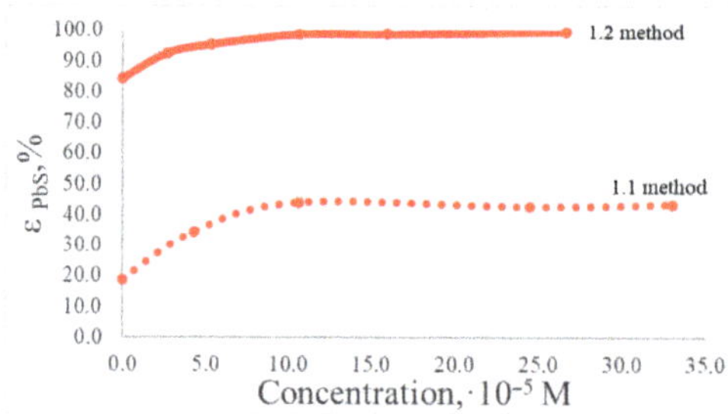

Figure 2. Dependence of PbS fraction (−71 + 41 μm) and (−41 + 10 μm) recovery with butyl xanthogenate using Hallimond tube microflotation (1.1 method) and froth flotation (1.2 method).

Flotation kinetics. Flotation kinetics were studied on narrow mineral fractions (−71 + 41 μm) of sulphides and Carbonaceous Material. The reagent addition sequence in the mechanical cell was as described above. Concentrate fractions were collected after 0.1, 0.5, 1, 2, 3 and 5 min.

The flotation rate constant (k, min^{-1}) for the rapidly floated fraction (1-min flotation) was determined graphically as the slope of the straight line in coordinates $\ln(100/(100-\varepsilon)) = f(t)$ (see Figure 3).

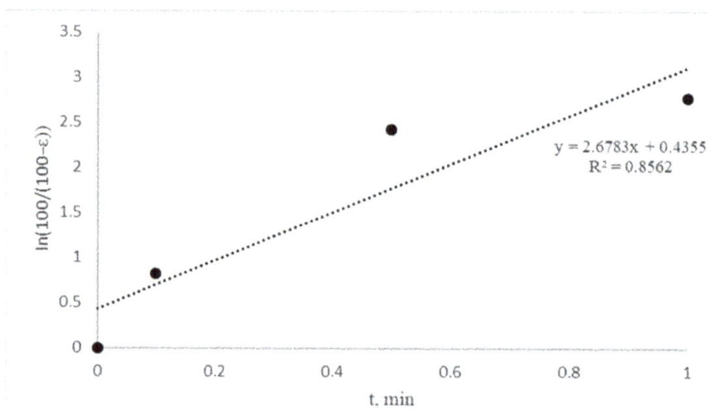

Figure 3. Determination of flotation rate constant (k, min^{-1}) for PbS with KButX (Fraction −71 + 41 μm, pretreated with NaOH, [KButX] = 10^{-4} M, pine oil; initial pH_{in} = 7.38 and Eh_{in}= +325 mV, final pH_{fin} = 7.8 and Eh_f = +312.5 mV).

Ore flotation. Ore flotation was conducted using two technological schemes. The first was direct-differential flotation at 80–85% passing −71 μm, with sphalerite depression using a combination of $ZnSO_4$ and $NaHSO_3$, galena recovery with KButX, followed by sphalerite activation with $CuSO_4$ in a lime medium and subsequent KButX flotation. This scheme included primary flotation, scavenging and two-stage re-cleaning for lead concentrate and three-stage re-cleaning for zinc concentrate, simulating a washing plant flowsheet. The second scheme was the proposed bulk-differential flotation with separation of CM at the process head (see Figure 4).

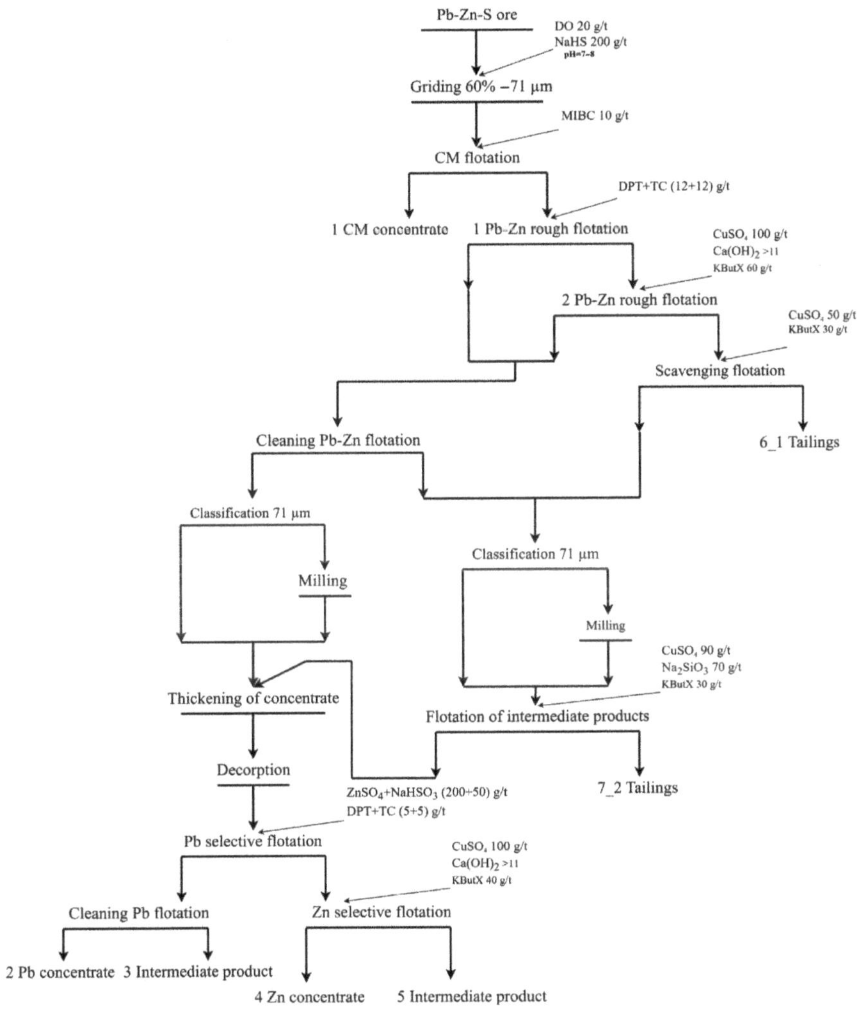

Figure 4. Laboratory bulk-differential flotation flowsheet.

Flotation efficiency was evaluated based on the selectivity of lead and zinc separation into selective saleable concentrates, expressed as the sum of lead and zinc recoveries:

$$\eta = \varepsilon PbPb + \varepsilon ZnZn \tag{2}$$

2.1.2. Electrokinetic Potential (Zeta Potential)

Zeta potential measurements of sulphides were carried out using a Malvern Nano ZEN 3600 analyzer. The particle size of ultra-fine sulphides was (2.6 ± 0.2) μm, with an S:L ratio of 1:2000. Measurements were performed in pH-adjusted water (using HCl or NaOH) in an electrochemical cell without added electrolyte. A 50 mg sample of the mineral was mixed with pH-adjusted water in a 100 mL volumetric flask (three parallel samples), then transferred to the electrochemical cell for analysis. The average value of three measurements was reported. Supporting electrolytes were not used, as sulphides possess sufficient intrinsic conductivity.

2.1.3. Adsorption

Sulphydric collectors adsorption. Adsorption of sulphydric collectors was determined as the difference between the initial (C_{in}, mol/L) and residual (C_{res}, mol/L) concentrations, taking into account the suspension volume (V, mL) and specific surface area (s·m, m^2).

$$A = \frac{V \cdot (C_{in} - C_{res})}{s \cdot m}, \ \text{mol/m}^2 \tag{3}$$

Relative adsorption (L, %) was calculated as follows:

$$L = \frac{C_{in} - C_{res}}{C_{in}} \cdot 100\% \tag{4}$$

Concentrations of sulphydric collectors were monitored using a UV-VIS-NIR Cary 600 spectrophotometer. Characteristic absorption bands for electronic $\pi \rightarrow \pi^*$ transitions in conjugated C(P)=S bonds were used: 301 nm (S–C=S, KButX, KEtX, solvent-water), 241 nm (N–C=S, TC, solvent-water), and 224 nm (P=S, DPT, solvent-hexane), with calibration software. Both initial and post-sorption solutions contained various forms of sulphydric collectors—ionic, disulphide and oxidation products. UV spectral analysis confirmed a decrease in all characteristic collector bands [22].

Adsorption studies were conducted on narrow fractions of sphalerite and galena (−41 +10 μm). Adsorption was performed under static conditions with stirring for 25 min, using 0.5 g aliquots and an S:L ratio of 1:100. Adsorption of sulphydric collectors was also monitored under microflotation conditions with the same aliquot mass and S:L ratio, with 25 min of pre-conditioning and flotation.

Acid-base centers. Acid-base active Brønsted centers on the surface within the studied pH range were identified using Gammet indicators [23–26]. Table 5 lists the characteristics of the Gammet indicators used to determine the types of acid-base centers present on the surface of the carbonaceous samples.

Table 5. Characteristics of Gammet indicators for acid-base center identification.

Indicator	Formula	Molecular Mass	λ, nm	pKa	Active Center Type
Bromocresol purple	$C_{21}H_{16}Br_2O_5S$	540.22	590	6.4	Bronsted acid
Bromthymol blue	$C_{27}H_{28}Br_2O_5S$	624.39	610	7.3	Bronsted base

A UV-VIS-NIR Cary 6000i spectrophotometer (Agilent Technologies, Santa Clara, CA, USA) with a wavelength range of 175 to 1800 nm was used to monitor the intensity of characteristic absorption bands before and after interaction with the carbonaceous sample.

3. Results and Discussion

3.1. Flotation

3.1.1. Microflotation in Hallimond Tube

Microflotation studies in the Hallimond tube provide indirect information on the comparative surface hydrophobicity of minerals, as frothing agents are absent from the reagent scheme. Preliminary alkaline treatment ensures equivalent surface oxidation rates for all samples. Four sulphydric collectors were tested on galena and sphalerite at an initial pH = 7. The results (Figure 5) show the dependence of floatability for PbS and ZnS with various sulphydric collectors at pH ≈ 7, fraction −44 + 10 μm, after NaOH pre-treatment.

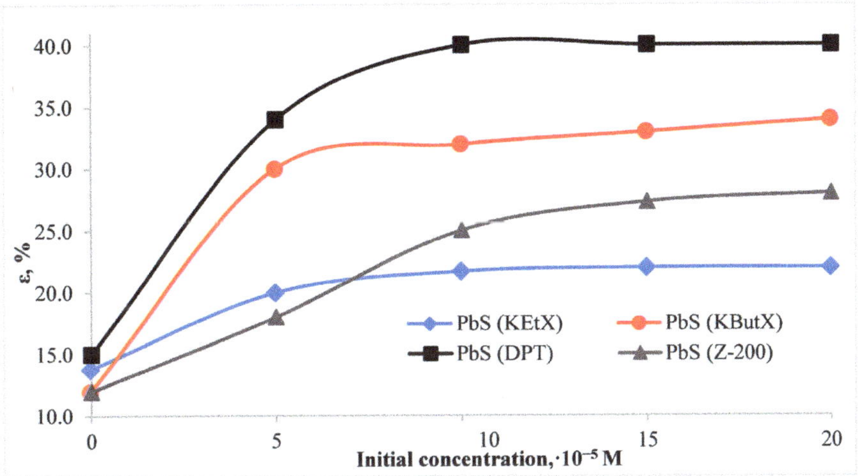

(**a**)

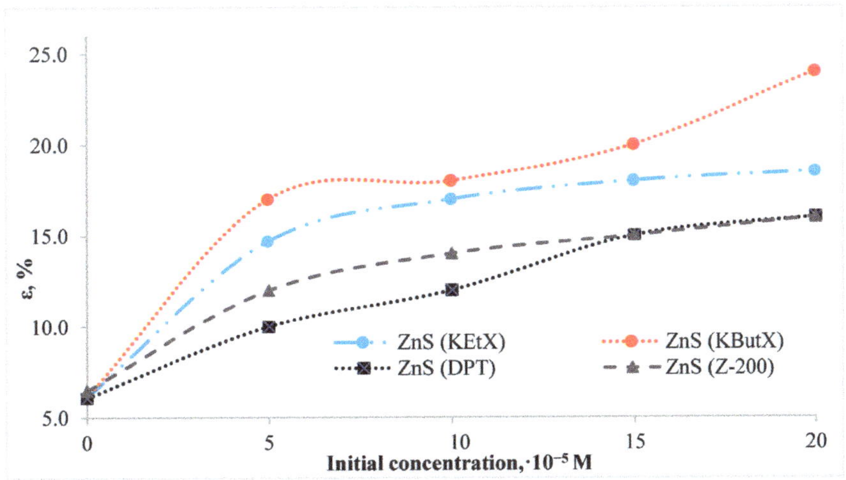

(**b**)

Figure 5. Dependents of sulphydric collectors (KEtX; KButX; DPT and TC (Z 200)) concentration on floatability of PbS (**a**) and ZnS (**b**) in Hallimond tube (pH ≈ 7; fraction −44 + 10 μm, with pre-treatment NaOH).

The flotation results indicate that galena exhibits higher flotation activity than sphalerite, consistent with previous findings [1]. For galena, the best results were obtained with dithiophosphate (DPT), which aligns with some literature data [27] but differs from the commonly reported highest floatability with xanthogenates. This discrepancy may be attributed to the use of molar concentrations (mol/L) rather than volumetric (mg/L) or the dose of a collector (g/t). For sphalerite, the best flotation was observed with butyl xanthogenate (ButX), while DPT provided the lowest recovery, highlighting the need for selective reagent schemes. Previous studies have shown that DPT exhibits low adsorption and collective ability toward sphalerite at low specific flow rates [28].

Analysis of DPT surface compounds using MAS NMR ˆ31P revealed bidentate-bridge coordination of dithiophosphate groups to two adjacent surface metal centers

(DPT:Zn = 1:2) for sphalerite and bidentate-terminal coordination (DPT:Pb = 1:1) for galena. Additional surface species identified include disulphides, thiophosphates, phosphates (hydrolysis products) and precipitated lead complexes [29,30].

3.1.2. Flotation Kinetics

Figure 6 presents the flotation kinetics for narrow monomineral fractions of sulphides and Carbonaceous Material (CM) under varying sulphide preparation conditions. Potassium butyl xanthogenate (KButX) was used as the sulphydric collector, and diesel oil (DO) as the apolar collector.

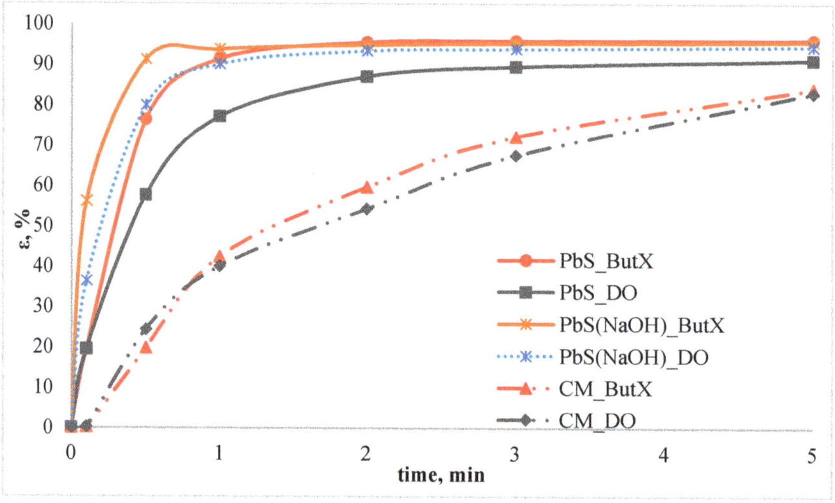

(**a**)

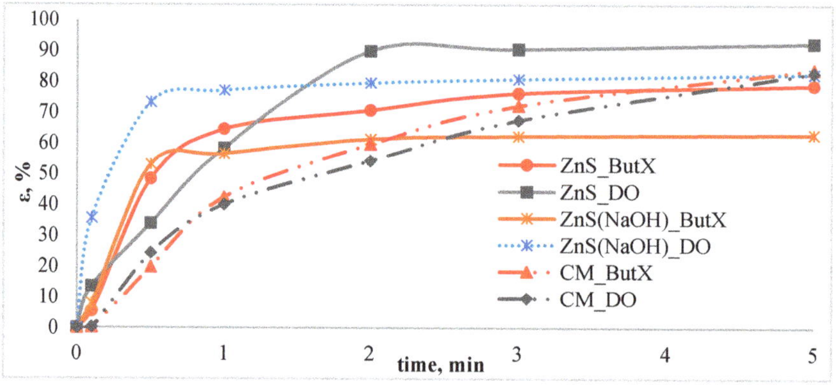

(**b**)

Figure 6. Flotation kinetics for PbS (**a**), ZnS (**b**) and CM (**a**,**b**) with KButX and DO with different sulphides preparation conditions; εmid ± 2% ([KButX] = 10^{-4} M (18 mg/L), [DO] = 10 mg/L, [P2] = 50 mg/L, with pre-treatment NaOH and without pre-treatment NaOH).

Preliminary alkaline treatment of sulphides (Figure 6a) significantly enhances the flotation activity of galena with KButX (ε_{PbS} = 91% in 0.5 min) compared with sphalerite (ε_{ZnS} = 61% in 2 min). Galena also demonstrates higher flotation activity with DO (ε_{PbS} = 90.1% in 1 min) than sphalerite (ε_{ZnS} = 77.1% in 1 min). These results are consistent with microflotation data obtained in the Hallimond tube.

When galena and sphalerite are tested with natural surface oxidation (mimicking aged sulphides, Figure 6b), both sulphides and CM reach similar recovery levels after 5 min of flotation (83–96%). For galena with KButX, the recovery plateau is achieved rapidly (ε_{PbS} = 96% in 0.5 min), while for sphalerite, the plateau is reached more slowly (ε_{ZnS} = 79% in 3 min). The apolar collector yields comparable recovery for galena and sphalerite after 5 min (ε_{PbS} = 89.5% in 2 min; ε_{ZnS} = 84%). CM is floated more slowly, but both collectors ultimately achieve similar recovery levels (ε_{CM} = 74% in 4 min). Notably, sphalerite exhibits higher floatability with DPT than with KButX. The floatability of sulphides and CM is comparable, especially for sulphides with natural oxidation, indicating that differential flotation under these conditions is challenging.

Processing of ores containing CM generally follows two approaches: preliminary flotation of the Carbonaceous Material, or its depression during sulphide flotation [8,12]. To depress Carbonaceous Material, reagents derived from wood processing are commonly used. Among these, reagent P2 has shown promise for depressing CM in gold-sulphide ores [31]. Figure 7 demonstrates flotation kinetics for PbS, ZnS and CM with KButX, and the effects of preliminary contact with the depressors P2 (Figure 7a) and LS (Figure 7b).

The use of depressor P2 (Figure 7a,b) substantially reduces the flotation activity of Carbonaceous Material, but also significantly retards sphalerite recovery, with maximum ZnS recovery not exceeding 20%. When DO is used as collector (Figure 7b), ZnS recovery does not exceed 50%, and CM recovery after 5 min is limited to 50%. In contrast, the lignosulphonate-based depressor LS has a less pronounced effect on sphalerite flotation, while galena flotation kinetics remain similar to those observed with collector alone. Notably, the flotation activity of CM recovers to 83% with KButX in the presence of LS (Figure 7a,b).

Table 6 summarizes the key flotation parameters for sulphides and CM after 1 min of flotation: the rate constant (k, min^{-1}) for rapidly floated grains, the linear trend approximation accuracy (R^2), recovery into concentrate after 1 min ($\varepsilon^{1'}$, %) and maximum recovery after 5 min ($\varepsilon_{max}{}^{5'}$, %).

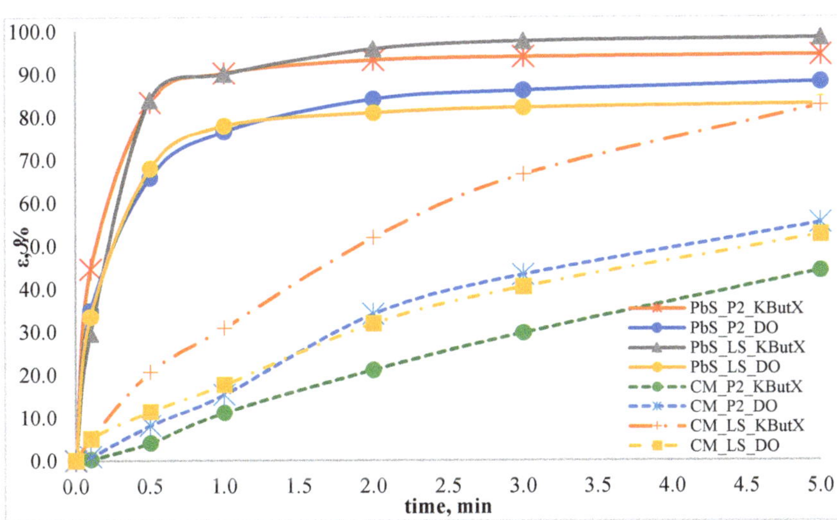

(a)

Figure 7. *Cont.*

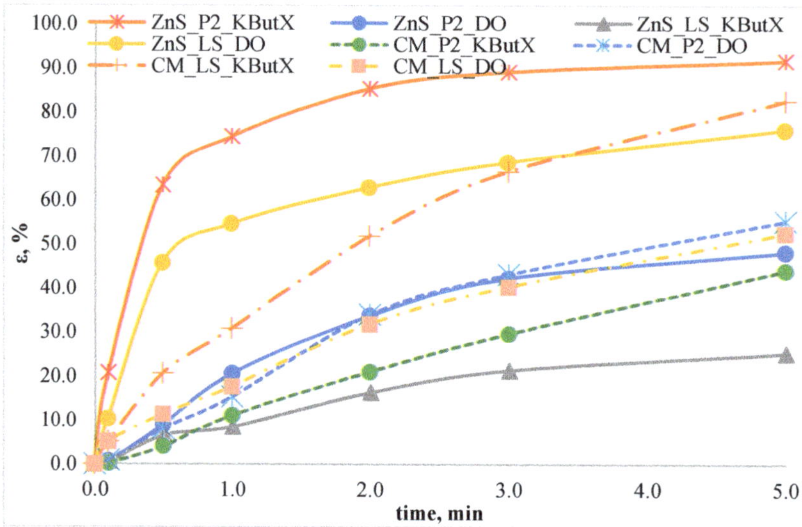

(**b**)

Figure 7. Flotation kinetics for PbS (**a**), ZnS (**b**) and CM with KButX and CM in presence of P2, and in presence of LS; $\varepsilon_{mid} \pm 2\%$. ([KButX] = 10^{-4} M (18 mg/L), [DO] = 10 mg/L, [depressor P2 or LS] = 50 mg/L or, without pre-treatment NaOH).

Table 6. Sulphides and CM flotation parameters after 1 min flotation (k, min^{-1}, $\varepsilon^{1'}$ and R^2 ($\varepsilon^{1'}$)) and ε_{max} after 1 min and 5 min flotation.

Reagent Scheme	k, min^{-1}	R^2 ($\varepsilon^{1'}$)	$\varepsilon^{1'}$ ($\varepsilon_{max}{}^{5'}$), %
Without pre-treatment NaOH (natural oxidation of the sulfide surface)			
PbS_KButX	2.46	0.96	91.4 (96.1)
ZnS_KButX	1.08	0.99	64.5 (78.9)
CM_KButX	0.61	1.00	42.4 (84.1)
PbS_KButX_LS	2.35	0.92	90.0 (98.4)
ZnS_KButX_LS	1.37	0.94	89.9 (91.8)
CM_KButX_LS	0.35	0.98	30.8 (82.7)
PbS_KButX_P2	2.23	0.92	90.4 (94.4)
ZnS_KButX_P2	0.10	0.92	8.5 (25.2)
CM_KButX_P2	0.13	1.00	11.1 (44.1)
PbS_DO	1.40	0.98	77.1 (91.1)
ZnS_DO	0.84	0.99	58.3 (92.5)
CM_DO	0.56	0.99	40.0 (83.0)
PbS_DO_LS	1.69	0.88	77.9 (83.0)
ZnS_DO_LS	0.74	0.90	54.6 (76.1)
CM_DO_LS	0.20	0.93	17.6 (52.5)
PbS_DO_P2	1.37	0.92	76.6 (88.2)
ZnS_DO_P2	0.25	0.99	20.6 (48.2)
CM_DO_P2	0.17	1.00	15.3 (55.3)
with NaOH and washing			
PbS_{NaOH}_KButX	2.68	0.86	93.8 (95.0)
ZnS_{NaOH}_KButX	1.12	0.81	56.6 (63.6)
PbS_{NaOH}_KButX_P2	1.89	0.92	90.0 (94.4)
ZnS_{NaOH}_KButX_P2	0.09	0.89	8.5 (25.2)
PbS_{NaOH}_DO	1.84	0.96	76.6 (88.2)
ZnS_{NaOH}_DO	0.81	0.76	56.6 (63.0)
PbS_{NaOH}_DO_P2	1.12	0.95	66.6 (84.2)
ZnS_{NaOH}_DO_P2	0.25	1.00	20.6 (48.3)

When lignosulphonate (LS)-based depressors are used in reagent schemes with naturally oxidized sulphides, the rate constant for rapidly floated galena grains decreases only slightly: from 2.46 min^{-1} (PbS_KButX) to 2.35 min^{-1} (PbS_KButX_LS) and from 1.40 min^{-1} (PbS_DO) to 1.69 min^{-1} (PbS_DO_LS). For sphalerite, the rate constant increases from 1.08 min^{-1} (ZnS_KButX) to 1.37 (ZnS_KButX_LS) min^{-1} but decreases from 0.84 min^{-1} (ZnS_DO) to 0.74 min^{-1} (ZnS_DO_LS). For Carbonaceous Material, the flotation rate constant decreases with both depressors: from 0.61 min^{-1} (CM_KButX) to 0.35 min^{-1} (CM_KButX_LS) and to 0.13 min^{-1} (CM_KButX_P2) and from 0.56 (CM_DO) to 0.20 min^{-1} (CM_DO_LS) and to 0.17 min^{-1} (CM_DO_P2). Despite this, the recovery of CM in the presence of both depressors increases linearly with flotation time. The preliminary removal of oxidized films significantly reduces the floatability of sphalerite under all reagent regimes. The absence of a clear contrast in the floatability of sphalerite, galena and Carbonaceous Material (CM) in the presence of LS and P2 depressants does not allow recommending the use of CM depression in the flotation reagent regime.

3.2. Electrokinetic Potential

The observed similarity in floatability between galena and sphalerite with naturally oxidized surfaces for both KButX and apolar DO collectors necessitated measurement of the electrokinetic potential (EKP) of sulphides. Figure 8 presents the dependence of EKP on the pH of the environment.

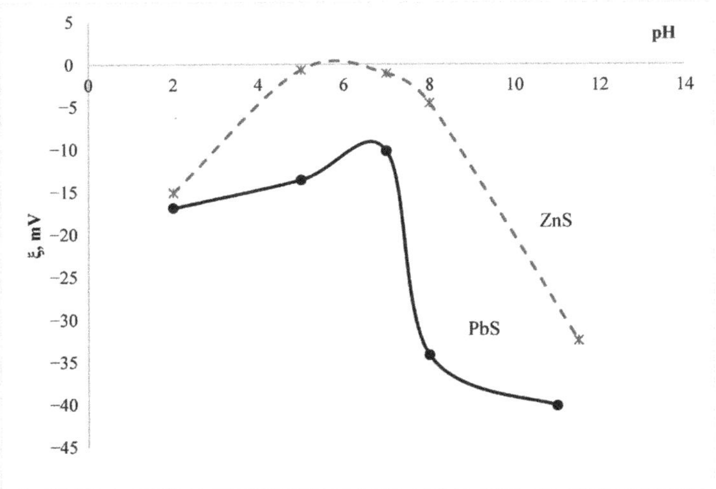

Figure 8. EKP (ξ) and pH correlation.

As shown in Figure 8, sphalerite exhibits EKP values near zero in the investigated pH range. This indicates a high potential for the attachment of non-ionogenic collectors and explains the effective flotation of sphalerite with diesel oil, which hydrophobizes the sphalerite surface and results in high recovery (approximately 63–83%, see Figure 6). The isoelectric point (IEP) for sphalerite is shifted from the low-acid region (pH 3.5–4) to a more neutral range (pH 4.5–7) compared with literature data. This shift is attributed to the absence of KNO_3 in the measurement procedure [32], consistent with previous methodological findings [33]. The IEP region is characterized by minimal electrostatic energy barriers for adsorption and maximum surface energy, facilitating the fixation of both ionogenic and non-ionogenic collectors.

The ability to hydrophobize the sphalerite surface with non-ionogenic collectors in the pH range optimal for galena flotation reduces the contrast in flotation response between galena and sphalerite.

3.3. Adsorption

3.3.1. Sulphydric Collectors Adsorption

Adsorption studies were carried out for both ionogenic and non-ionogenic collectors to identify a sulphydric collector with the greatest adsorption contrast. Figure 9 shows the adsorption isotherms for galena and sphalerite under conditions similar to those used in microflotation tests with the Hallimond tube (see Figure 5).

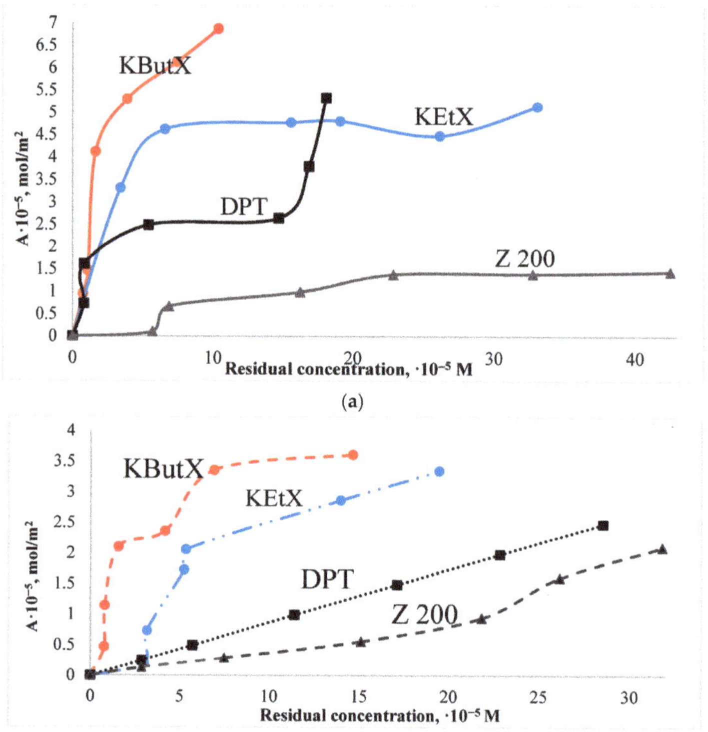

Figure 9. Collectors adsorption dependences for PbS (**a**) and ZnS (**b**) (pH ≈ 7; fraction −44 + 10 μm, with pre-treatment NaOH).

As shown in Figure 9, the adsorption of sulphydric collectors on galena is up to twice as high as on sphalerite. Most sulphydric collectors exhibit polymolecular adsorption. Xanthogenates and dithiophosphates (at low initial and residual DPT concentrations) display the highest affinity for galena, while butyl xanthogenate is most strongly adsorbed on sphalerite. Dithiophosphate (DPT) provides the greatest adsorption contrast between galena (A_{PbS} = 1.1 mol/m^2) and sphalerite (A_{ZnS} = 0.49 mol/m^2) at the same initial concentration, which can be exploited for differential flotation, particularly at low DPT concentrations [29,30].

Figure 10 summarizes the relationship among zeta potential (ζ, mV), relative adsorption of KButX (L, %), and sulphide recovery in Hallimond tube flotation (ε, %) at pH 8 (adjusted with NaOH).

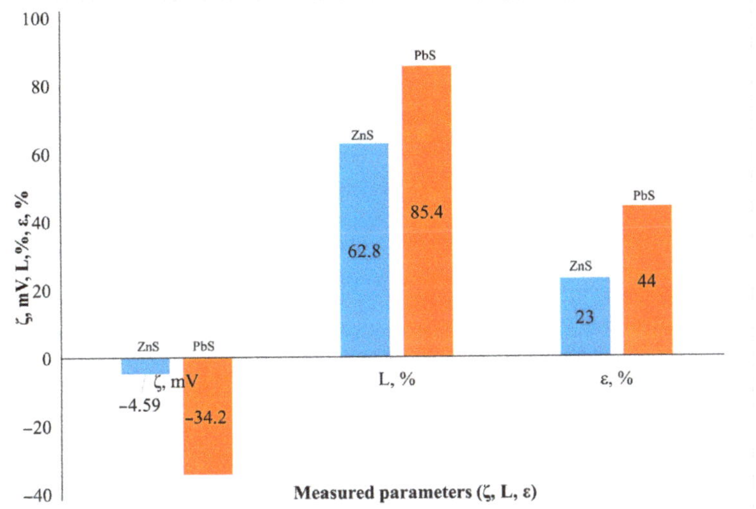

Figure 10. Comparative data diagram of ZnS and PbS: recovery (ε, %) into concentrate; relative adsorption (L, %) KButX under flotation conditions in Hallimond tube (Size −44 + 10 μm, pH_{inNaOH} = 8.0, [KButX] = 10^{-4} M) and Zeta Potential (ξ, mV) of ultra-fine grain (pH_{inNaOH} = 8.0, [KButX] = 10^{-4} M).

At pH 8, KButX provides higher galena recovery (ε_{PbS} = 44%) compared with sphalerite (ε_{ZnS} = 23%), which is consistent with higher relative adsorption (L_{PbS} = 85.4% vs. L_{ZnS} = 62.8%). The highest galena recovery corresponds to the most negative zeta potential, indicating that KButX is adsorbed primarily in ionic form on galena, while both ionic and molecular forms can adsorb on sphalerite at this pH.

3.3.2. Acid-Base Centers

Table 7 presents the results of acid-base center determination within the pH range of 6–8 during flotation and adsorption.

Table 7. Results of finding acid-base centers within a pH range of 6–8.

W/o NaOH	Bronsted Acid pKa = 6.4 μmol/m^2	Bronsted Base pKa = 7.3 μmol/m^2
PbS	3.0	25.3
ZnS	4.3	10.0
CM	0.3	4.4

Within the studied pH range, the predominant surface centers are Brønsted bases, which act as electron donors in donor-acceptor interactions with reagents. This suggests that surface reactions can occur with non-ionogenic collectors containing unsaturated bonds and O, N, or P atoms capable of donor-acceptor interactions. Additionally, Brønsted acid-base centers on PbS and ZnS surfaces may interact with lignosulphonate-based depressors via dative hydrogen bonds, which can disrupt flotation contrast and selectivity.

CM is a mixed-phase association, consisting mainly of carbonates and layered aluminosilicates coated with carbon films, imparting hydrophobicity to the surface. According to flotation kinetics (Figures 6 and 7, Table 6), the flotation activity of CM increases proportionally with flotation duration.

The established comparable flotation activity of low-impurity sphalerite (cleiophane) in low-alkaline environments, along with the influence of natural oxidation of galena and sphalerite surfaces on flotation activity, justifies the use of a bulk-differential flotation technology for lead-zinc sulphide ores. To minimize the recovery of lead and zinc sulphides into the CM concentrate, NaHS is added as a depressor to the ground fraction. Table 8 presents the results of flotation tests on polymetallic ore containing cleiophane-type sphalerite and Carbonaceous Material, comparing bulk-differential and direct-differential flotation flowsheets. The collective cycle employs KButX, while the differential cycle uses minor amounts of DPT to maximize adsorption and flotation contrast between galena and sphalerite.

Table 8. Metallurgical balance of the components of laboratory flotation flowsheet test with ore in open mode.

	Product	γ, %	β, %			ε, %			Experiment Conditions
			Pb	Zn	C	Pb	Zn	C	
					Bulk-differential flotation flowsheet (Figure 4)				
1	CM concentrate	1.90	0.41	2.13	18.23	0.7	1.1	33.7	
2	Pb concentrate	2.80	30.80	6.14	1.62	77.9	4.5	4.4	
3	Intermediate product 2	2.27	4.21	2.37	0.78	8.6	1.4	1.7	grinding to 60% < 71 μm, DO 6 g/t and NaHS
4	Zn concentrate	5.72	0.69	53.59	1.19	3.6	79.7	6.6	carbonaceous flotation
5	Intermediate product 5	2.41	0.53	8.67	1.08	1.2	5.4	2.5	bulk Pb-Zn flotation
	Pb-Zn concentrate	13.2	7.65	26.52	1.19	91.3	91.0	15.3	
6	1 Tailings	74.50	0.08	0.35	0.55	5.4	6.8	39.9	desorption
7	2 Tailings	10.40	0.28	0.45	1.10	2.6	1.2	11.1	selection $ZnSO_4$ + $NaHSO_3$
	Final tailings	84.9	0.10	0.36	0.62	8.0	8.0	51.0	
	Ore	100.0	1.11	3.85	1.03	100.0	100.0	100.0	
					Direct-differential flotation flowsheet				
1	Pb concentrate	2.6	22.56	12.85	21.74	50.7	8.9	50.2	grinding to 85% < 71 μm $ZnSO_4$ + $NaHSO_3$
2	Zn concentrate	4.6	3.81	53.94	7.21	15.4	67.4	29.9	
	Final tailings	92.8	0.42	0.94	0.24	33.9	23.7	20.0	Pb flotation, with sphalerite depression
	Ore	100.0	1.14	3.68	1.11	100.0	100.0	100.0	Zn flotation, after activation by copper sulfate in high-alkaline lime environment.

The end products of the bulk flotation cycle are Carbonaceous Material, a bulk concentrate and final tailings. As shown in Table 8, the selectivity index (η) for direct-differential flotation is 132.8, while for bulk-differential flotation it is 118.1. The increased flotation activity of cleiophane accounts for the lower technological performance in the direct-differential scheme. The bulk-differential flotation flowsheet for the Shalkiya deposit polymetallic ore achieves a significant reduction in lead and zinc losses to final tailings. Furthermore, the flowsheet and reagent regimes in the bulk concentrate selection cycle can be further optimized.

4. Conclusions

Experimental results demonstrate that low-iron sphalerite (cleiophane) with natural surface oxidation can be recovered by both sulphydryl (KButX) and apolar (DO) collectors, similar to galena. The ability of sphalerite to be recovered with diesel oil in the pH range 6–8 is attributed to its Zeta Potential being close to 0 mV in this interval. Preliminary alkaline treatment of monomineralic fractions of galena and sphalerite prior to flotation, resulting in faster flotation of galena compared to sphalerite.

Dithiophosphate (DPT) provides a threefold difference in floatability between galena and sphalerite at a concentration of 10^{-4} mol/L, corresponding to lower static adsorption on sphalerite (A_{ZnS} = 0.49 mol/m^2, comparable to A_{PbS} = 1.1 mol/m^2 at the same initial concentration). Therefore, DPT is recommended for use in the lead-zinc concentrate selection stage or as part of a direct-differential flotation flowsheet.

Within the studied pH range (6–8), galena, sphalerite, and CM surfaces are dominated by Brønsted base sites, which reduces the flotation contrast between galena and sphalerite and complicates the depression of CM by lignin-sulphonates.

Kinetic studies of galena, sphalerite, and CM flotation with butyl xanthogenate and diesel oil, in the presence of CM depressors P2 and LS, show a reduction in the flotation rate constant for both CM and sulphides, particularly sphalerite. The flotation kinetics of CM increase linearly with flotation time, indicating that selective depression of CM relative to galena and sphalerite is not achievable under these conditions.

The results obtained for monomineral fractions of cleiophane, galena and CM provide a scientific basis for the application of a bulk-differential flotation flowsheet with preliminary recovery of CM for the beneficiation of complex lead-zinc sulphide ores from the Shalkiya deposit.

Author Contributions: Conceptualization, R.T., M.N. and V.I.; methodology, V.I.; validation, M.K., A.Y. and N.Y.; investigation, A.Y., M.K., N.Y. and A.A.; resources, R.T. and M.K.; data curation, R.T. and V.I.; writing—original draft preparation, A.Y., V.I., N.Y. and A.A.; writing—review and editing, T.K., R.T. and A.Y.; supervision—M.N.; project administration, R.T. and T.K.; funding acquisition, R.T. All authors have read and agreed to the published version of the manuscript.

Funding: This research was funded by the Science Committee of the Ministry of Science and Higher Education of the Republic of Kazakhstan, grant number AP19680419.

Data Availability Statement: The original contributions presented in this study are included in the article. Further inquiries can be directed to the corresponding author.

Conflicts of Interest: The authors declare no conflicts of interest.

References

1. Sutherland, K.L.; Wark, I.W. *Principles of Flotation*, Rev. and enlarged ed.; Australasian Institute of Mining and Metallurgy: Melbourne, Australia, 1955; 489p.
2. Ignatkina, V.A.; Bocharov, V.A. Features of flotation of various copper sulfides and sphalerite contained in sulfide ore. *Gorn. Zhurnal* **2014**, *12*, 75–79. (In Russian)
3. Flores, L.F.; Montoya, A.; Ramírez, A.; Aguilar, J.; Reyes, I.A. Flotation of non-activated sphalerite using xanthogen formate-type compounds as collector reagents: Adsorption study and optimization. *Colloids Surf. A Physicochem. Eng. Asp.* **2025**, *714*, 136586. [CrossRef]
4. Ralston, J.; Alabaster, P.; Healy, T.W. Activation of zinc sulphide with CuII, CdII and PbII: III. The mass-spectrometric determination of elemental sulphur. *Int. J. Miner.* **1981**, *7*, 279–310. [CrossRef]
5. Hu, Y.; Wu, M.; Liu, R.; Sun, W. A review on the electrochemistry of galena flotation. *Miner. Eng.* **2020**, *150*, 106272. [CrossRef]
6. Nayak, A.; Jena, M.S.; Mandre, N.R. Beneficiation of lead-zinc ores—A review. *Miner. Process. Extr. Metall. Rev.* **2022**, *43*, 564–583. [CrossRef]
7. Ralston, J. The chemistry of galena flotation: Principles and practice. *Miner. Eng.* **1994**, *7*, 715–735. [CrossRef]
8. Pan, Z.; Xiong, J.; Cui, Y.; Wei, Q.; Jia, W.; Zhang, Z.; Jiao, F.; Qin, W. Effect mechanism of carbonaceous materials on the flotation separation of lead–zinc ore. *Sep. Purif. Technol.* **2022**, *294*, 121101. [CrossRef]
9. Glembotsky, V.A.; Dmitrieva, G.M. *Vliyanie Genezisa Mineralov na ikh Flotatsionnye Svoistva (Influence of Mineral Genesis on Their Flotation Properties)*; Nauka: Moscow, Russia, 1965; 292p. (In Russian)
10. McFadzean, B.; Castelyn, D.G.; O'Connor, C.T. The effect of mixed thiol collectors on the flotation of galena. *Miner. Eng.* **2012**, *36–38*, 211–218. [CrossRef]
11. Peng, Y.; Grano, S.; Fornasiero, D.; Ralston, J. Control of grinding conditions in the flotation of galena and its separation from pyrite. *Int. J. Miner. Process.* **2003**, *70*, 67–82. [CrossRef]
12. Gredelj, S.; Zanin, M.; Grano, S.R. Selective flotation of carbon in the Pb–Zn carbonaceous sulphide ores of Century Mine Zinifex. *Miner. Eng.* **2009**, *22*, 279–288. [CrossRef]
13. Kostovic', M.; Gligoric, Z. Multi-criteria decision making for collector selection in the flotation of lead–zinc sulfide ore. *Miner. Eng.* **2015**, *74*, 142–149. [CrossRef]

14. Hua, Z.; Shi, B.; Dong, Y.; Fu, Y.; Zeng, Y.; Sun, W.; Liu, R.; Tang, H. Evaluation of Mineralogical Characteristics and Flowsheet Improvements for Carbon-Bearing Lead-Zinc Ore. *JOM* **2025**, *77*, 830–841. [CrossRef]
15. Zhao, J.; Hu, W.; Xiao, F.; Liu, X.; Yu, H.; Yuan, H.; Wang, H. Characterization of discarded lead–zinc sulfide ore tailings based on mineral fragments. *Minerals* **2022**, *12*, 1279. [CrossRef]
16. Konieczny, A.; Pawlos, W.; Krzeminska, M.; Kaleta, R.; Kurzydlo, P. Evaluation of organic carbon separation from copper ore by pre-flotation. *Physicochem. Probl. Miner. Process.* **2013**, *49*, 189–201. [CrossRef]
17. Sime, F.M.; Jin, J.; Wang, X.; Wick, C.D.; Miller, J.D. Characterization and simulation of graphite edge surfaces for the analysis of carbonaceous material separation from sulfide ores by flotation. *Miner. Eng.* **2022**, *182*, 107590. [CrossRef]
18. Chimonyo, W.; Fletcher, B.; Peng, Y. Starch chemical modification for selective flotation of copper sulphide minerals from carbonaceous material: A critical review. *Miner. Eng.* **2020**, *156*, 106522. [CrossRef]
19. Chimonyo, W.; Peng, Y. Review of the Influence of Native and Modified Biopolymers on Carbonaceous Gangue Depression in Selective Flotation. *Miner. Process. Extr. Metall. Rev.* **2024**, *46*, 593–610. [CrossRef]
20. Lee, S.; Gibson, C.; Borschneck, A.; Ghahreman, A. Transformer oil vs. kerosene: Selective collectors for C-matter flotation from a double refractory gold ore. *Miner. Eng.* **2023**, *191*, 107951. [CrossRef]
21. Yushina, T.I.; Yergeshev, A.R.; Dumov, A.M.; Makavetskas, A.R. Study of the material composition of lead-zinc ore of the Shalkiya deposit in order to determine the possibility of its processing. *Non-Ferr. Met.* **2022**, *53*, 8–14. [CrossRef]
22. Grano, S.R.; Prestidge, C.A.; Ralston, J. Sulphite modification of galena surfaces and its effect on flotation and xanthate adsorption. *Int. J. Miner. Process.* **1997**, *52*, 1–29. [CrossRef]
23. Tanabe, K. *Solid Acids and Bases: Their Catalytic Properties*, 1st ed.; Academic Press: New York, NY, USA, 1970; 184p.
24. Nechiporenko, A.P. Donor-acceptor properties of the surface of solid-phase systems. In *Indicator Method: Tutorial*; Lan: St. Petersburg, Russia, 2017; 284p, ISBN 978-5-8114-2309-5. (In Russian)
25. Atchabarova, A.A.; Abdimomyn, S.K.; Abduakhytova, D.A.; Kishibayev, K.K.; Zlobina, Y.V.; Kurbatov, A.P.; Turdean, G.L.; Djenizian, T. Electrochemical modification of the carbon material surface by hydroxyl groups. *J. Solid State Electrochem.* **2024**, *28*, 2425–2436. [CrossRef]
26. Ryazantseva, M.V.; Bunin, I.Z. Modifying acid–base surface properties of calcite, fluorite and scheelite under electromagnetic pulse treatment. *J. Min. Sci.* **2015**, *51*, 1016–1020. [CrossRef]
27. Zhang, L.; Gao, J.; Khoso, S.A.; Wang, L.; Liu, Y.; Ge, P.; Tian, M.; Sun, W. A reagent scheme for galena/sphalerite flotation separation: Insights from first-principles calculations. *Miner. Eng.* **2021**, *167*, 106885. [CrossRef]
28. Ignatkina, V.A.; Bocharov, V.A.; D'yachkov, F.G. Enhancing the disparity in flotation properties of nonferrous metal sulfides using sulfhydryl collecting agents with different molecular structures. *J. Min. Sci.* **2015**, *51*, 389–397. [CrossRef]
29. Ivanov, A.V.; Antzutkin, O.N.; Larsson, A.; Kritikos, M.; Forsling, W. Polycrystalline and surface O,O%-dialkyldithiophosphate zinc(II) complexes: Preparation, 31P CP:MAS NMR and single-crystal X-ray diffraction studies. *Inorganica Chim. Acta* **2001**, *315*, 26–35. [CrossRef]
30. Larsson, A.-C.; Ivanov, A.V.; Antzutkin, O.N.; Forsling, W. A 31P CP/MAS NMR study O,O′-dialkyldithiophosphate lead (II) complexes. *J. Colloid Interface Sci.* **2008**, *327*, 370–376. [CrossRef]
31. Sosipatorov, A.I.; Panchenko, G.M.; Vysotin, V.V.; Vinokurova, M.A.; Chikin, A.Y. Application prospects of domestic depressor reagent under carbonaceous gold-bearing ore flotation. *Vestn. Irkutsk. Gos. Tekhnicheskogo Univ. (Proc. Irkutsk. State Tech. Univ.)* **2018**, *22*, 184–193. (In Russian) [CrossRef]
32. Albrecht, T.W.J.; Addai-Mensah, J.; Fornasiero, D. Critical copper concentration in sphalerite flotation: Effect of temperature and collector. *Int. J. Miner. Process.* **2016**, *146*, 15–22. [CrossRef]
33. Ignatkina, V.A.; Bocharov, V.A.; Aksenova, D.D.; Kayumov, A.A. Zeta potential of the surface of ultrafine sulfides and floatability of minerals. *Russ. J. Non-Ferr. Met.* **2017**, *58*, 95–100. [CrossRef]

minerals

MDPI

Article

Composition, Properties, and Flotation Reagent Regimes of Carbonaceous Material

Akim Yergeshev [1,2], Rustam Tokpayev [1,*], Marina Karmeeva [2], Tamina Khavaza [1,*], Mikhail Nauryzbayev [1] and Vladislava Ignatkina [2]

1 Centre of Physico-Chemical Methods of Research and Analysis, Al-Farabi Kazakh National University, Almaty 050000, Kazakhstan; akim9797@mail.ru (A.Y.); nauryzbaev@cfhma.kz (M.N.)

2 Department of Mineral Processing and Industrial Wastes Recycling, College of Mining, National University of Science and Technology "MISIS" (NUST MISIS), Moscow 119049, Russia; m2001044@edu.misis.ru (M.K.); woda@mail.ru (V.I.)

* Correspondence: rustamtokpaev@mail.ru (R.T.); k.tamina@mail.ru (T.K.); Tel.: +7-707-522-44-88 (R.T.); +7-747-280-93-98 (T.K.)

Abstract

The novelty of this study lies in the first comparative characterisation of five carbonaceous materials: three monophase carbons (wood charcoal, carbon, graphite) and two ore-derived CM samples from polymetallic sulphide and oxidised lithium ores. The methodology included IR spectroscopy, XPS, acid–base adsorption centres identified by colour indicators, chemical composition analysis, and kinetic flotation tests. Bulk and surface compositions differed significantly: although the ash content of ore-derived CM reached 84.4%, XPS revealed carbon-enriched surfaces with thin films of about 1–2 nm. IR spectra confirmed multiphase structures with carbonate, silica, and aluminosilicate bands, and showed an identical composition of CM from different industrial ore types. Flotation kinetics confirmed high floatability (recoveries 80%–99%, k up to 1.95 min^{-1}). Even with sodium lignosulphonate at 500 mg/L, recovery only decreased from 83.02% to 52.54%, showing the limited efficiency of depressants. These results provide a basis for the preliminary removal of CM prior to rough (bulk) flotation in the processing of different ore types, improving concentrate quality, reducing reagent consumption, and lowering metallurgical losses.

Keywords: carbonaceous material (CM); IR spectra; XPS; acid–base centres of adsorption by the colour of the indicator; froth flotation; flotation rate constant; apolar collector; depressor

1. Introduction

CM is commonly found in various types of ores, and its presence negatively impacts the technological performance of flotation and cyanidation processes [1–3]. Processing CM is a critical challenge in mineral processing, as it can be either recovered by flotation into a froth concentrate (similar to apolar substances and minerals), depressed during flotation, or subjected to decomposition by microorganisms [4–8].

Previously, it was established [3] that CM is multicomponent in nature, consisting of charcoal and graphite remains, along with quartz and dolomite inclusions. Microscopic studies have revealed that CM has a multiphase composition, including graphite, amorphous carbon (such as kerogen), and other mineral inclusions [2,8–10].

In sulphide ores, natural CM often contains significant graphite structures with varying crystalline grain sizes, orientations, levels of disorder, and surface functional groups [11–13].

 https://doi.org/10.3390/min15090974

Surface studies using modelling and instrumental techniques have shown that the face surface of graphite exhibits natural hydrophobicity, while the edge surface contains oxidised functional groups such as C-O-C and C=O [14–16].

In the flotation of sulphide ores, butyl or amyl xanthates are commonly used as collectors. These compounds oxidise to form dixanthogenide molecules, which enhance CM floatability [1,17–22]. Furthermore, CM, due to its high floatability, can be floated using a frother alone [23–25]. Despite this activity, efforts to depress CM using polymolecular compounds of either natural (carbohydrate derivatives) or synthetic origin (polyethyleneoxides and polyacrylamides) have not yielded sustainable depressant effects [26–28].

Lee S. et al. [3] highlighted the controversy between the amount of fixed apolar collectors and the flotation rate. For simpler ores containing CM, kerosene has been shown to achieve higher recovery rates and faster flotation as compared to transformer oil. IR-Fourier spectroscopy analysis revealed stronger interactions between transformer oil and coal surfaces compared to kerosene.

The relationship between apolar reagent viscosity and flotation performance was addressed in the monograph [29], stating that the higher-viscosity apolar collectors require lower reagent consumption but longer mixing times before flotation.

Despite extensive research efforts, approaches to CM processing via flotation, whether by recovery into concentrate or tailings, remain inconsistent. However, the negative impact of CM on metallurgical balance during flotation or cyanidation processes has been unequivocally recognised. The present work has identified causal relationships between the composition of carbon-containing samples, their surface properties, and their flotation behaviour.

2. Materials and Methods

2.1. Samples

Five carbon-bearing samples were studied, including three monophase substances: wooden charcoal (sample 1W_C), obtained by pyrolysis for household use; carbon (sample 2A_C), used as an adsorbent for organic floating reagents during desorption operations prior to bulk flotation concentrate selection; and graphite (sample 3G_C), a monophasic mineral that forms the base of CM and determines CM floatability [16]. Two additional samples were derived from non-sulphide lithium micaceous ore (sample 4Li_C) from the Shavazsay deposit in Uzbekistan and polymetallic sulphide ore (sample 5S_C) from the Shalkiya deposit in Kazakhstan.

The three monophase carbon-bearing materials were manually crushed using an agate mortar, and the narrow particle-size fraction (41–71 μm) was separated using a mechanical sieve set for subsequent analyses of elemental composition, specific surface area, and adsorption properties. In contrast, the two carbonaceous samples derived from ore deposits (sample 4Li_C from the Shavazsay deposit in Uzbekistan and sample 5S_C from the Shalkiya deposit in Kazakhstan) were prepared by grinding averaged ore batches (10 kg each) in a Knelson Gravity Solutions rod mill at a ratio of S:R:W = 1:6:0.7, until 35%–40% of the material passed 71 μm. The entire ground product was then air-dried at room temperature, disaggregated in a mortar, and subjected to gravitational separation in a spiral separator SL-500 with a water flow rate of 3 L/min. The light fraction was floated at natural alkalinity without flotation reagents, and the froth product was again air-dried. After homogenisation in an agate mortar, the narrow size fraction (41–71 μm) was separated by mechanical sieving for further investigation.

The elemental composition of the carbonaceous samples was determined using a combination of instrumental methods. C, H, N, and S elements were analysed with a Leco CHN-628 analyser (LECO Corporation, St. Joseph, MI, USA, while oxygen content was measured using the GOST2408.3-90 method [30]. Non-organic ash elements in Li_C and S_C samples were analysed using the MLA Quanta650 system (FEI Company, Hillsboro, OR, USA), which includes a scanning electron microscope (SEM Quanta650) equipped with an energy-dispersive detector EDAX Silicon Drift controlled by Genesis software integrated into MLA Suite. Specific surface area measurements were performed using the BET method on a SORBTOMETR analyser by ZAO Katakon, Novosibirsk, Russia. Ash content was determined according to the GOST11022-95 method, with samples burned in a muffle furnace at 840 °C [31].

The chemical composition and specific surface area of the studied sample fractions are presented in Table 1:

Table 1. Chemical composition and specific surface area ± standard deviation, %.

	Sample	C ±2.0	H ±3.1	N ±0.2	O ±0.5	S ±0.2	Ash Content A, %, ±0.5	Specific Surface Area, m^2/g, ±2.00
1	Wood charcoal (W_C)	81.1	4.1	0.3	9.9	0.2	4.4	10.34
2	Carbon (A_C)	84.0	5.0	1.5	7.0	0.3	2.2	318.10
3	Graphite (G_C)	98.0	-	-	1.0	-	1.0	9.98
4	Li mica (Li_C)	2.5	-	1.0	6.0	0.1	90.5	23.36
5	Pb-Zn-S ore (S_C)	10.2	-	-	4.7	0.7	84.4	12.93

The highest carbon content was observed in graphite (G_C), which contained 98% of carbon. Carbon (A_C) exhibited the largest specific surface area, measured at 318.1 m^2/g. The specific surface area values for carbonaceous samples separated from ores (S_C and Li_C) were significantly higher than those typically found in similar narrow fractions of sulphides, which are generally much less than 1 m^2/g, and rock minerals, which tend to be slightly more than 1 m^2/g.

CM samples separated from ores (Li_C and S_C) showed notably high ash contents, with values of 90.5% for Li_C and 84.4% for S_C. Table 2 presents the elemental composition of ash obtained after burning these CM samples using the EDAX Silicon DriftSEM Quanta 650 system.

Table 2. Elemental composition of CM ash, %.

CM	O ±2.0	Si ±1.0	Mg ±0.6	Al ±0.2	Ca ±3.0	Zn ±2.3	Fe ±2.6	K ±0.6	Na ±0.6	S ±1.7
Li_C	65.1	10.1	0.8	5.8	6.7	-	5.2	1.7	0.9	3.7
S_C	56.9	5.1	13.3	0.3	5.0	5.2	4.6	-	3.6	6.0

Lead (Pb) was not detected in the ash composition of the S_C CM (Table 2), consistent with the analytical methodology outlined in the reference. Electron microscopy confirmed the multiphase nature of the CM and its heterogeneous mineral composition, with ash components consisting of oxide, silicate, and sulphate mineral forms.

2.2. Reagents of Flotation

The study employed diesel oil (apolar collector) and butyl xanthate (ionogenic sulphhydryl collector), with pine oil as the frother. Ligninosulphonate (Polyplast Ligno) was used as a depressant.

Butyl xanthate was selected for the tests as it is a typical sulphhydryl collecting agent commonly used in the reagent regimes of sulphide and complex ores. A thiol collector (potassium butyl xanthate) served as the conventional sulphide collector and benchmark [32].

An apolar collector (diesel oil) was included to represent the hydrophobisation of graphitic CM [33].

Polyplast Ligno (PL) was used as a depressor to reduce the flotation activity of the carbonaceous samples [34]. This depressor is a modified liquid lignosulphonate product, appearing as a brown liquid with 50% activity. Lignosulphonates are derivatives of lignin and byproducts of wood processing, consisting of high-molecular compounds with molecular masses starting from 200 g/mol. In this experiment, solutions were prepared assuming 100% activity.

In general, sodium lignosulphonate (LS) was tested with each collector as a low-toxicity depressant; prior studies document LS adsorption/depression on MoS_2 and on xanthate-treated chalcopyrite with pronounced pH/lime dependence [35,36], and its greener use in Cu–Mo separation with collector-dependent effects on chalcopyrite [37]; and selective wettability effects in LS/frother media further support LS applicability [38].

2.3. Research Methods

2.3.1. IR Spectroscopy

The molecular composition of carbonaceous samples was analysed using a two-ray spectrophotometer (Specord M-80, Carl Zeiss, Jena, Germany) equipped with an Eberth monochromator and diffraction grating for spectral decomposition. IR spectra were processed using Soft Spectra software. Samples were ground in an agate mortar, mixed with two drops of butanol, and applied onto KBr glass. Measurements were conducted in the 4000–400 cm^{-1} range with three integrations (IT = 3) in transmission mode (T, %).

2.3.2. X-Ray Photoelectron Spectroscopy Analysis

X-ray photoelectron spectroscopy (XPS) was performed on the 41–71 μm fraction of samples to determine the elemental composition of the surface layer (2 nm depth) and assess oxygen/carbon states in coal and graphite. Analyses were conducted using a PHI5000 VersaProbeII spectrometer (ULVAC-PHI, ULVAC-PHI Inc., Chigasaki, Japan) with monochromatized Al Kα radiation (hν = 1486.6 eV, 50 W power, 200 μm beam diameter). Integral intensities were measured for core-level lines: C 1s, O 1s, Si 2s, Mg 1s, Al 2p, Ca 2p, Zn 2p, F 1s, Pb 4d5, Fe 2p, K 2p, Na 1s, S 2s, F 1s, and N 1s. High-resolution spectra for C 1s and O 1s were acquired at 23.5 eV analyser transmission energy with a 0.2 eV step. Peak fitting used a nonlinear least-squares method with Gaussian–Lorentzian functions, including asymmetry correction for sp^2 carbon in C 1s peaks. The binding energy (E_B) scale was calibrated using Au 4f (83.96 eV) and Cu 2p3 (932.62 eV).

2.3.3. Acid–Base Centres

For the direct determination of Lewis–Brønsted active acid–base centres on the surface, colour (Gammet) indicators with different pKa values were employed [39–42]. According to Tanabe, Lewis base centres (electron donors) contain unshared electron pairs and are capable of capturing protons during the hydrolytic dissociation of water molecules. This is demonstrated by the selective adsorption of Gammet indicators with negative pKa values. Brønsted acid centres are characterised by the selective adsorption of indicators with pKa values between 1 and 7, which is attributed to surface OH-groups. The presence of Brønsted base centres is indicated by the adsorption of indicators with pKa values between 7 and 14. Lewis acid centres correspond to the selective adsorption of indicators with pKa values ≥ 14.

Table 3 provides the characteristics of the Gammet indicators used to identify the types of acid–base centres on the surface of the studied carbonaceous samples based on their adsorption behaviour.

Table 3. Gammet indicators' characteristics to define types of acid–base centres.

Indicator	Formula	Molecular Mass	λ, nm	pKa	Active Centre Type
Dinitroaniline	$C_6H_5N_3O_4$	183.12	340	−4.4	Lewis base
Brilliant green	$C_{27}H_{34}N_2O_4S$	482.64	610	1.3	
Bromphenol blue	$C_{19}H_{10}Br_4O_5S$	669.97	590	4.1	Brønsted acid
Bromocresol purple	$C_{21}H_{16}Br_2O_5S$	540.22	590	6.4	
Bromthymol blue	$C_{27}H_{28}Br_2O_5S$	624.39	430	7.3	Brønsted base
Indigo carmine	$C_{16}H_8N_2Na_2O_8S_2$	466.36	610	12.8	
M-Dinitrobenzene	$C_6H_4N_2O_4$	168.11	240	16.8	Lewis acid

Indicator solutions were prepared at a concentration of 10^{-4} mol/L in a water–ethanol solution (60:40). Adsorption studies were conducted using carbonaceous samples from the −71 + 41 µm fraction.

The quantitative test for determining the dominant type of acid–base centre was based on changes in the absorption intensity of characteristic indicator bands, corresponding to specific pKa values, after mixing with carbonaceous samples for two hours. A blank sample was prepared by conditioning the carbonaceous sample with distilled water for two hours, followed by careful decantation of water and subsequent addition of an indicator solution with the same pKa value to create an equivalent indicator concentration. The absorption intensity of characteristic bands was then measured for comparison. The concentration of specific surface-active centres, equivalent to the amount of adsorbed colour indicator (g_{pKa}), was calculated as follows:

$$g_{pK_a} = \frac{C \cdot V}{I_0} \cdot \left(\left| \frac{/I_0 \cdot I'/}{m \cdot s} \pm \frac{/I_0 \cdot I_{bl.s}/}{m_{bl.s} \cdot s} \right| \right) \quad (1)$$

g_{pK_a}—amount of indicator adsorbed on sample surface, mol/m^2;
C—indicator solution concentration, mol/mL;
V—indicator solution volume, ml;
s—carbonaceous sample specific surface area, m^2/g;
m, $m_{bl.s}$—mass of sample weight in main and blank parts, g;
I_0, I', $I_{bl.s}$—intensity of initial indicator solution adsorption after interaction with carbonaceous sample in blank experiment, respectively, Abs.

A UV-VIS-NIR Cary 6000i spectrophotometer (Agilent Technologies, Santa Clara, CA, USA) with a wavelength range of 175–1800 nm was used to monitor changes in absorption intensity before and after interaction with carbonaceous samples.

Figure 1 illustrates an example of determining the adsorption intensity for CM from lead–zinc ore. As shown in Figure 1, the adsorption intensity of the indigo carmine indicator increased after contact with CM from ore, indicating electronic transition activation during interaction between the indicator and acid–base centres on the S_C surface.

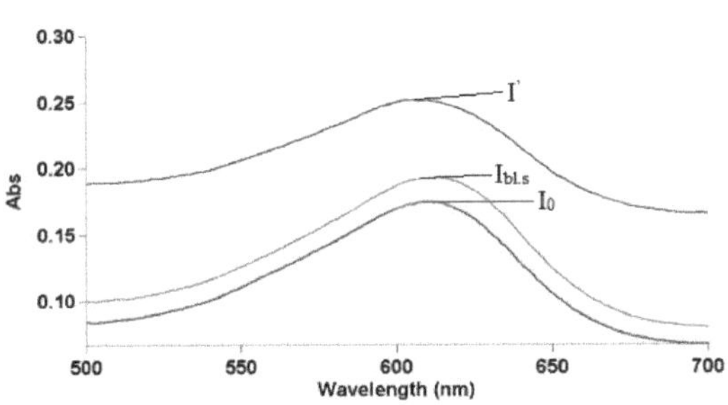

Figure 1. Example of defining characteristic strip absorption intensity for indigo carmine on S_C.

2.3.4. Froth Flotation

The flotation kinetics of the samples were studied using a specific fraction of 41–71 μm and a sample weight of 5 g in a laboratory mechanical flotation machine, FML-0.3 (Mekhanobr, St. Petersburg, Russia), equipped with a 50 mL chamber. In the mechanical cell, suspension mixing and air dispersion were generated by an impeller, and a foam removal device was used to collect the froth. The flotation reagent regime (Table 4) was conducted under natural environmental conditions created by distilled water. The flotation reagents included diesel oil, butyl xanthate, and pine oil as the frother.

Table 4. Carbonaceous samples reagent flotation regime.

Flotation Agent	Frother–Pine Oil	Diesel Oil and Frother	Butyl Xanthate and Frother	Depressor (PL)
Concentration, mg/L	10	20 and 10	20 and 10	50 and 500

Table 4 presents the details of the reagent flotation regime.

In the flotation cell, the carbonaceous sample was mixed with distilled water for 2 min, followed by sequential mixing with the collecting agent for 1 min and the frother for another minute. When lignosulphonate was used as a depressor for carbonaceous samples, an additional 2 min of mixing was performed before adding the collecting agent.

After specific time intervals, froth fractions were skimmed into separate cups at 0.1 min (1 stroke of froth scraper), 0.5, 1, 2, 3, and 5 min. The concentrate fractions and tails were dried and weighed, and the material balance of the experiment was calculated. The concentrate yield in each fraction corresponded to recovery (ε), assuming α = β = 100%. The flotation rate constant (k) was calculated similarly to the rate constant of a first-order chemical reaction, where the reaction product is represented by a "particle-air bubble" flotation complex:

$$\varepsilon = \varepsilon_{max}(1 - e^{-kt}) \quad (2)$$

k is the rate constant (min^{-1}), ε_{max} is the maximum recovery, and t is the flotation time (min).

Figure 2 illustrates an example of graphite (G_C) flotation kinetics using only a frother (left) and its transformed dependence in semilogarithmic coordinates $\ln \frac{100}{100-\varepsilon} = f(t)$ (right).

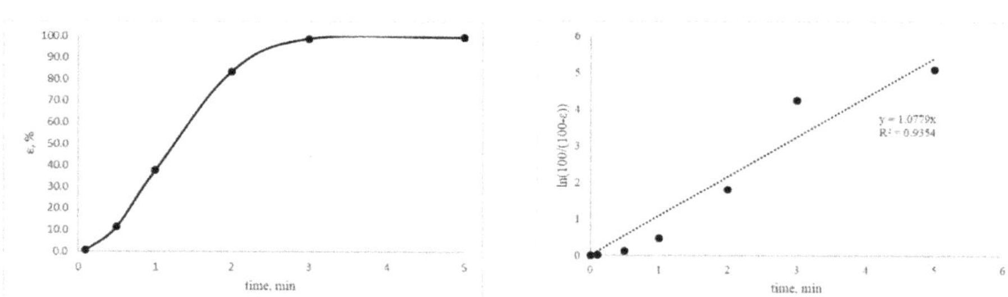

Figure 2. G_C graphite flotation kinetics with frother (**left**) and graphic definition k, min^{-1} (**right**).

The flotation rate constant (k) was determined from the linear equation y = ax, where ln(ε) = kt (Figure 2, right). In this example of graphite flotation with only a frother, the calculated rate constant was k = 1.17 min^{-1}. To ensure the reliability and reproducibility of the results, each flotation kinetics experiment was repeated three times. The average values of these experimental results were used for data analysis.

3. Results and Discussion

3.1. IR Spectroscopy

Figure 3 presents the overall IR spectra of wooden charcoal, carbon (W_C, A_C), graphite (G_C) (Figure 3a), and carbonaceous samples from Pb-Zn sulphide ore (S_C) and Li micaceous ore (Li_C) (Figure 3b).

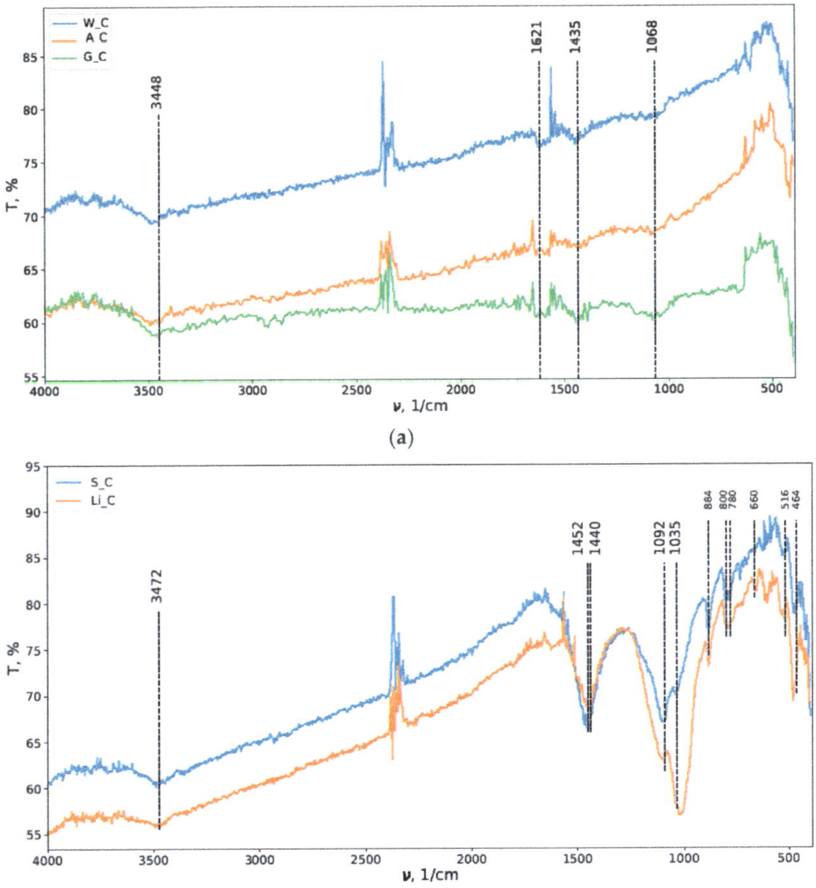

Figure 3. IR spectra of carbon-containing samples: (**a**) monophased (W_C, A_C, G_C), (**b**) multi-phased CM.

The IR spectra of the monophase samples (Figure 3a), including wooden charcoal (W_C), carbon (A_C), and graphite (G_C), exhibit stretching vibrations of bound water (ν_{OH}) at 3448 cm^{-1} and deformation vibrations (δ_{OH}) at 1621 cm^{-1}. Additionally, stretching vibrations of the C-O bond (ν_{CO}) are observed at 1435 cm^{-1}, along with ν_{CO} at 1068 cm^{-1}, indicating the presence of oxidised centres within the structure of coal and graphite.

Of particular interest is the similarity in the characteristic bands of the CMs S_C and Li_C (Figure 3b), which were recovered from sulphide and non-sulphide ores, respectively. IR spectroscopy reveals that multiphase CM from various types of ores can be identified within the wave number range of 1600–400 cm^{-1}. This includes carbonate (C-O stretching vibrations at ν_{CO} = 1452 cm^{-1} and deformation vibrations at 884 cm—^{1}), silica (Si-O bond vibrations at 1008 cm^{-1} and out-of-plane vibrations at 464 cm^{-1}), and aluminosilicates

(Si-O bond vibrations at 1092 and 1035 cm^{-1}, deformation vibrations at 880 and 780 cm^{-1}, and out-of-plane vibrations at 516 cm^{-1}). These bands correspond to X-O bonds, where X can be Si, C, or other elements present in the carbonaceous sample.

The presence of multiple bands within the range of 1570–700 cm^{-1} suggests a distorted tetrahedral structure. A broader characteristic band with maximum transmission at 3472 cm^{-1} corresponds to O-H bond stretching vibrations (ν_{OH}), while deformation vibration bands δ_{OH} within the range of 1620–1600 cm^{-1} indicate the presence of bound water.

Despite the presence of bound water and hydrophilic components, such as silica, in carbonaceous samples S_C and Li_C, these materials are naturally hydrophobic and exhibit high flotation activity. This behaviour highlights their unique structural properties, which contribute to their flotation performance.

3.2. XPS

The elemental composition of the surface of the studied samples was analysed using XPS (Table 5).

Table 5. Sample surface elemental composition, ± standard deviation, %.

	C ±1.0	O ±1.0	Si ±0.5	Mg ±0.5	Al ±0.5	Ca ±0.5	P ±0.1	N ±0.3	F ±0.5	Zn ±0.5	Fe ±0.3	K ±0.1	Na ±0.1	S ±0.2	Pb ±0.1
W_C	87.8	10.2	-	-	-	0.7	0.1	0.9	-	-	-	0.3	-	-	-
A_C	85.1	12.0	1.4	-	0.8	0.7	-	-	-	-	-	-	-	-	-
G_C	98.5	1.5	-	-	-	-	-	-	-	-	-	-	-	-	-
Li_C	18.9	49.3	12.9	5.9	4.2	2.3	-	1.1	3.6	-	0.7	1.0	0.1	-	-
S_C	49.2	34.5	5.1	2.5	1.9	2.5	-	-	-	2.8	0.4	-	-	0.7	0.4

Comparing the elemental composition of the samples (Table 5), it is evident that the carbon concentration on the surface of CMs from Li_C (18.9%) and S_C (49.2%) is significantly lower than that of other carbonaceous samples (85.1%–98.5%). In contrast, the oxygen concentration is higher for Li_C (49.3%) and S_C (34%) compared to other samples, which range from 1.5% to 12.0%. Additionally, noticeable concentrations of elements such as Si, Mg, Al, Ca, and Zn are observed in these samples.

Graphite (G_C) is identified as the purest sample among those studied. The C 1s spectrum of graphite consists of an asymmetrical narrow peak (E_B = 284.4 eV, FWHM = 0.69 eV) and a wide π-π^* satellite shifted by 6.2 eV, which is characteristic of sp^2 hybridisation in graphite atoms. In contrast, the A_C sample contains carbon associated with sp^3 hybridised atoms (non-crystalline carbon) and functional groups bonded with oxygen.

The C 1s spectrum of W_C differs significantly from that of graphite. Its approximation reveals two main peaks: an asymmetrical peak at 284.9 eV and a symmetrical peak at 287.2 eV, along with a third peak at 289.9 eV. The location of Peak 1 differs from similar peaks in G_C and A_C, and asymmetry is observed in this peak as well. The distance between Peaks 1 and 2 is 2.3 eV, suggesting that Peak 2 may correspond to epoxy groups. Peak 3 is attributed to carbon atoms in specific functional groups:

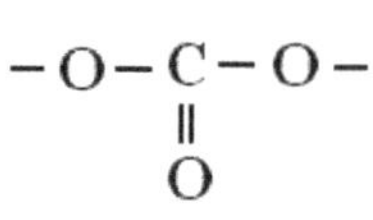

Figure 4 presents the O 1s spectra for activated carbon W_C (left) and graphite G_C (right), highlighting their differences.

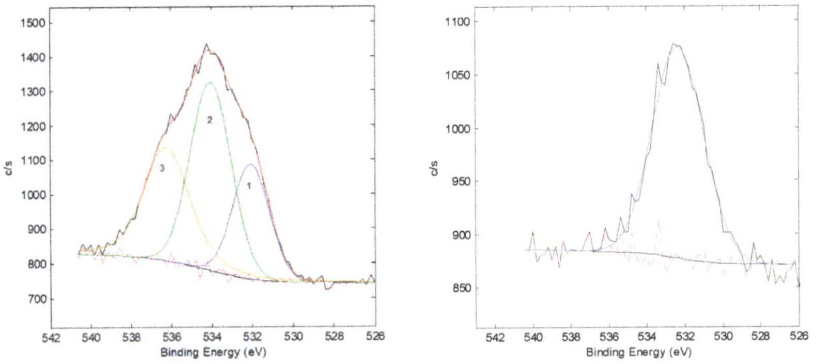

Figure 4. O 1s XPS scans of W_C (**left**) and G_C (**right**); 1 (≈ 532.0 eV)—C–O; 2 (≈ 534.0 eV)—adsorbed H_2O and/or strongly H-bonded –OH; 3 (≈ 536.0 eV)—above the typical O 1s range for carbon surfaces, attributed to differential charging; a contribution from interlayer molecular O_2 cannot be ruled out but requires verification (mild UHV bake-out and/or charging control).

The analysis of the O 1s spectrum for W_C shows a complex form with three distinct peaks at 532.0 eV, 534.0 eV, and 536.0 eV. The binding energy (E_B) of the first peak corresponds to adsorbed oxygen, while the second peak can be associated with adsorbed water molecules. The binding energy of Peak 3 exceeds the typical upper limit for O 1s spectrum values, which may be explained by differential charge effects or molecular oxygen trapped between carbon layers.

The composition of CM from ores (Li_C and S_C) is notably more complex than that from coal (W_C), carbon (A_C), or graphite (G_C). Ash content determination results (Table 1) indicate predominantly non-organic content in Li_C (A = 90.5%) and S_C (A = 84.4%). The discrepancy between chemical analysis results and XPS surface data (Table 5) suggests that carbon adheres to the surfaces of silicates and other rock minerals.

The IR spectra of Li_C and S_C samples (Figure 3b) exhibit common features such as Si-O bond vibration bands within the range of 1100–800 cm^{-1}, indicative of silicates, and C-O vibration bands within the ranges of 1500–1350 cm^{-1} and 800–780 cm^{-1}, characteristic of carbonate minerals.

These findings suggest the formation of tight associations between silicate rock mineral surfaces and monophase carbon, which possesses naturally hydrophobic properties. This association contributes to the flotation activity observed in multiphase mineral assemblages (CM) during flotation and explains the challenges in depressing CM effectively.

It can be hypothesised that reducing ore grinding fineness prior to flotation would lead to homogenisation of CM within the pulp, making selective froth recovery or selective depression of carbonaceous substances impractical.

3.3. Acid–Base Centres

XPS data confirm the presence of O, C, and additional elements such as N (in W_C and Li_C), F (in Li_C), and S (in S_C) on the surface of all studied samples. These elements can act as active donor–acceptor centres, facilitating the fixation of flotation agents on the surface. IR spectra further reveal the presence of chemical and hydrogen bonds involving carbon, silicon, and hydrogen with oxygen (C-O, Si-O, and O-H) in the samples. Table 6 presents the results of experimental determination of active acid–base centres at respective acidity constants (pKa).

Table 6. Data overview on amount of active Brønsted–Lewis centres on carbonaceous samples' surface at different pKa, ± standard deviation %.

Indicator	Active Centre Type	pKa	g_{pKa}, μmol/m^2, ±0.25%				
			W_C	A_C	G_C	Li_C	S_C
Dinitroaniline	Lewis base	−4.4	2.45	0.77	0.08	0.24	3.84
Brilliant green	Brønsted acid	1.3	0.68	0.06	0.33	0.04	0.21
Bromphenol blue	Brønsted acid	4.1	0.06	0.17	0.02	0.13	0.37
Bromocresol purple	Brønsted acid	6.4	0.97	3.87	0.70	0.54	0.27
Bromthymol blue	Brønsted base	7.3	0.31	1.30	0.36	0.68	4.40
Indigo carmine	Brønsted base	12.8	1.95	0.41	0.53	3.63	31.34
M-Dinitrobenzene	Lewis acid	16.8	1.61	0.33	0.21	3.86	4.27

The highest concentration of active centres on the surface of carbonaceous samples was observed for material from sulphide ore (S_C). This sample exhibited Brønsted-type base centres at pKa = 12.8 and pKa = 7.3, with concentrations of g_{pKa} = 31.34 μmol/m^2 and g_{pKa} = 4.40 μmol/m^2, respectively. Additionally, S_C showed active Lewis base centres at pKa = −4.4 with a concentration of g_{pKa} = 3.84 μmol/m^2, as well as Lewis acid centres at pKa = 16.8 in the amount of g_{pKa} = 4.27 μmol/m^2.

For CM extracted from non-sulphide ore (Li_C), the highest concentration of active centres (g_{pKa} = 3.86 μmol/m^2) corresponded to Lewis-type acid centres at pKa = 12.8. In wooden charcoal (W_C), Lewis base centres at pKa = −4.4 were found in a concentration of g_{pKa} = 2.45 μmol/m^2, while Brønsted base centres at pKa = 12.8 were present in a concentration of g_{pKa} = 1.95 μmol/m^2.

Activated carbon (A_C) exhibited Brønsted-type acid active centres at pKa = 6.4 with a concentration of g_{pKa} = 3.87 μmol/m^2, while graphite (G_C) showed a lower concentration of similar centres at g_{pKa} = 0.70 μmol/m^2. Among all samples, graphite demonstrated the lowest adsorption of indicators with different pKa values, aligning with its low impurity content as determined by XPS (Table 5). However, the presence of O-H hydrogen bonds was confirmed for graphite through IR spectroscopy (Figure 2, left).

Quantitative adsorption of Gammet indicators across different pKa values highlights the diversity of acid–base centres on the surface of carbonaceous samples. The predominant centres are Brønsted and Lewis bases, which act as potential electron donors in donor-acceptor interactions, including those involving water dipoles. These findings are consistent with IR spectroscopy results indicating bound water presence and XPS data identifying water within the layers.

Thus, the surface of CM samples contains active centres capable of fixing flotation agents through various donor–acceptor mechanisms via intermolecular interactions or complex formation. CM sourced from ores can interact with different types of surface-active compounds (collectors and frothers), reducing the contrast in surface technological properties during flotation processes.

3.4. Flotation Kinetics

Mao, L., and Yoon, R.-H [43] detailed that the flotation rate constant reflects the combined response of all subprocesses in the elementary flotation act—collision, fixation, and conservation.

In the flotation kinetics tests conducted on narrow fractions of carbonaceous samples using the studied reagent schemes, a linear relationship was observed in semilogarithmic coordinates. Table 7 summarises the flotation parameters, including the flotation rate constant k (min^{-1}), R-squared value (R^2), and overall carbonaceous samples recovery ε_{max}.

Table 7. Experiment results on flotation kinetics.

Sample		Flotation Parameters k, $min^{-1}/R^2/\varepsilon_{max}$				
		Pine Oil	Diesel Oil + Pine Oil	ButX + Pineoil	PL + Diesel Oil + Pine Oil	
					50 mg/L	500 mg/L
1	W_C	0.53/0.90/93.51	0.44/0.98/92.37	0.46/0.99/90.57	0.69/0.95/98.00	0.50/0.87/95.7
2	A_C	0.23/0.94/68.43	0.27/0.90/77.67	0.29/0.99/83.92	0.26/1.00/72.83	0.24/1.00/69.50
3	G_C	1.08/0.94/99.39	1.95/0.93/99.98	1.21/0.88/99.40	0.94/0.92/99.39	0.84/0.94/98.98
4	Li_C	0.18/0.96/61.10	0.30/1.00/78.57	0.32/0.97/79.50	0.23/0.99/68.29	0.17/0.99/56.53
5	S_C	0.15/0.87/81.15	0.33/0.96/83.02	0.36/0.98/84.07	0.26/1.00/72.00	0.14/0.99/52.54

As shown in Table 7, graphite (G_C) and wooden charcoal (W_C) exhibit the highest flotation activity based on their flotation rate constants and maximum recovery within 5 min of the process. The presence of the sulphhydryl collector ButX in the reagent scheme enhances flotation parameters for all samples except graphite (G_C), where it decreases from 1.95 min^{-1} (diesel oil collector) to 1.21 min^{-1} (ButX collector). However, graphite maintains high froth product recovery (ε_{max}) at 99%–100%. Preliminary mixing of graphite with PL depressor at two concentrations reduces the flotation rate constant to 0.94 min^{-1} (50 mg/L) and 0.84 min^{-1} (500 mg/L), while maximum extraction into froth concentrate remains at 99%.

Table 8 provides data on pH measurements in the final tailings under different reagent schemes.

Table 8. pH values at studied reagent schemes.

Reagent Scheme	$pH_{initial}$	pH_{final}				
		W_C	A_C	G_C	Li_C	S_C
pine oil	5.6 ± 0.4	6.20	7.30	5.40	8.56	5.53
diesel oil + pine oil	5.6 ± 0.4	6.20	7.25	5.70	8.61	5.66
ButX + pineoil	5.6 ± 0.4	5.96	7.40	5.76	8.56	5.84

For most samples (Li_C, A_C, and W_C), there is a typical shift toward alkaline pH due to hydrolysis of base centres. However, for polymetallic sulphide ore-derived CM (S_C), a shift toward acidic pH is observed, likely due to the presence of sulphide sulphur on the surface, which lowers pulp pH through hydrolysis. Graphite (G_C) exhibits minimal pH change during flotation, consistent with its purity and low concentration of acid–base centres on the surface (Table 5).

Figure 5 illustrates two diagrams: one showing the flotation rate constants for carbonaceous samples under different reagent schemes (Figure 5a) and another depicting the impact of PL concentration on k (min^{-1}) (Figure 5b) and εmax using diesel oil as a collector (Figure 5c).

The obtained flotation rate constants indicate that graphite has the highest k, consistent with its low impurity content and minimal indicator adsorption (g_{pKa} = 0.70 μmol/m^2) at pKa = 6.4 in comparison with other CMs (Table 5). The high flotation rate constant for graphite when using apolar collectors is attributed to its perfect edge cleavage and strong adhesion to apolar collectors. When butyl xanthate is used as a collector (k = 1.21 min^{-1}), its flotation performance is comparable to that achieved with only a frother (k = 1.17 min^{-1}). For other samples, k min^{-1} values are significantly lower, ranging from 0.15 to 0.46 min^{-1}, with maximum recovery during 5 min varying between 61.10% and 93.51%. CMs derived from ores contain hydrophilic silicates but exhibit maximum recovery levels around 80% due to carbon adhesion on their surfaces. Higher surface carbon concentrations—49.2%

for S_C compared to 18.9% for Li_C—correlate with higher froth product recovery (ε_{max} = 81%–84%) for S_C versus (ε_{max} = 61%–80%) for Li_C.

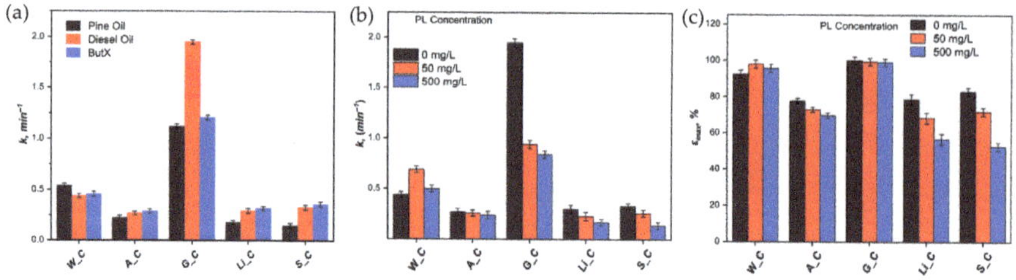

Figure 5. Diagram of flotation rate constant (**a**) k, min^{-1} of carbonaceous samples at different reagent schemes and influence of depressor PL on (**b**) flotation rate constant $k \cdot 10^2$ and (**c**) ε_{max}, %.

The use of butyl xanthate in reagent schemes increases ε_{max} for carbonaceous samples of various origins due to multiple active centres on their surfaces interacting with collectors and molecular dixanthogenide adhesion. However, a decrease in k is noted for graphite under these conditions.

The application of PL as depressors before adding collectors reduces graphite's flotation rate constant by approximately half (Figure 5b), while maintaining high maximum extraction at 99% (Figure 5c). Conversely, depressor use increases both k and maximum extraction for wooden charcoal.

In current flotation practice, diverse industrial ore types—including sulphide (including precious-metal-bearing), oxidised, and complex polymetallic ores—commonly contain CM. The combined evidence from IR spectroscopy, XPS, the acid–base indicator method for surface centres, and froth-flotation tests in this study shows that selective depression of CM is not a reliable primary solution. CM retains high floatability and is only partially depressed; its presence in saleable concentrates complicates downstream metallurgy by lowering the grade of target-metal concentrates, increasing slag burden and reagent consumption, and, in hydrometallurgical routes, enabling preg-robbing of metal complexes [2,19,44,45]. These observations are consistent with our recent findings [46], which further demonstrated that CM remains highly floatable even under depressing conditions.

Based on these findings, it is recommended to prioritise preliminary extraction of CM through coarse grinding from ores using apolar collectors during flotation washing of sulphide and non-sulphide ores rather than relying on depressors for CM depression.

4. Conclusions

Chemical composition analysis, ash content measurements, specific surface area studies, and XPS and IR spectroscopy reveal that CM from two distinct industrial ores is multiphase in nature. These ores predominantly contain silicate and carbonate minerals, which exhibit strong surface coalescence with carbon. This compositional feature influences the natural hydrophobic behaviour and adsorption capacity of the material. IR spectroscopy effectively identifies CM through characteristic bands in the wavenumber range of 1600–400 cm^{-1}.

Brønsted solid acid–base centres are the most prevalent active sites for CM within the pKa range corresponding to flotation pH conditions. The high reactivity of CM is attributed to oxygen centres in silicate and carbonate tetrahedra, which interact via proton transfer, combined with the natural hydrophobicity provided by carbon on the surface.

Flotation kinetics studies demonstrate that graphite exhibits the highest floatability in the presence of an apolar collector (k = 1.97min^{-1}), primarily due to adhesion facilitated by its perfect cleavage edges. Other carbon-containing samples show flotation rate constants ranging from 0.20 to 0.46 min^{-1} across all reagent regimes studied, including butyl xanthate as a collector, with maximum recovery into froth products reaching 80%–94%. The floatability of CMs is strongly correlated with the carbon concentration in their surface layers.

PL derivatives used as depressors for wood charcoal increased its flotation rate constant from 0.23 min^{-1} to 0.50–0.69 min^{-1}, depending on depressor concentration in the liquid phase, while maintaining high recovery, approximately 96%–98%. For other carbon-containing samples, PL slightly reduced flotation rate constants but preserved a high maximum recovery to concentrates (over 50%).

The recovery of CM using apolar collectors within flotation reagent regimes for sulphide and non-sulphide ores is demonstrated to be a more rational and effective approach compared to relying on depressors for reducing floatability.

Author Contributions: Conceptualisation, V.I., R.T., and M.N.; methodology, V.I.; validation, M.K., A.Y.; investigation, A.Y., M.K., and T.K.; resources, R.T. and M.N.; data curation, V.I.; writing—original draft preparation, A.Y., V.I., and T.K.; writing—review and editing, A.Y., V.I., T.K., and R.T.; supervision, M.N.; project administration, R.T. and T.K.; funding acquisition, R.T. All authors have read and agreed to the published version of the manuscript.

Funding: This research was funded by the Science Committee of the Ministry of Science and Higher Education of the Republic of Kazakhstan, grant number AP19680419.

Data Availability Statement: Data is contained within the article.

Conflicts of Interest: The authors declare no conflicts of interest.

References

1. Sousa, R.; Futuro, A.; Pires, C.S.; Leite, M.M. Froth flotation of Aljustrel sulphide complex ore. *Physicochem. Probl. Miner. Process.* **2017**, *53*, 758. [CrossRef]
2. Adams, M.D.; Burger, A.M. Characterization and blinding of carbonaceous preg-robbers in gold ores. *Miner. Eng.* **1998**, *11*, 919. [CrossRef]
3. Lee, S.; Gibson, C.; Borschneck, A.; Ghahreman, A. Transformer oil vs. kerosene: Selective collectors for C-matter flotation from a double refractory gold ore. *Miner. Eng.* **2023**, *191*, 107951. [CrossRef]
4. Tabatabaei, R.H.; Nagaraj, D.R.; Vianna, S.M.; Napier-Munn, T.J.; Gorain, B. The effect of non-sulphide gangue minerals on the flotation of sulphide minerals from Carlin-type gold ores. *Miner. Eng.* **2014**, *60*, 26. [CrossRef]
5. Wieniewski, A.; Skorupska, B. Technology of Polish copper ore beneficiation—Perspectives from the past experience. *E3S Web Conf.* **2016**, *8*, 01064. [CrossRef]
6. Kowalczuk, P.B.; Zaleska, E.; Danczak, O. Flotation of carbonaceous copper shale–quartz mixture with poly(ethylene glycol) alkyl ethers. *Trans. Nonferr. Met. Soc. China* **2015**, *25*, 314. [CrossRef]
7. Magwaneng, R.S.; Haga, K.; Batnasan, A.; Shibayama, A.; Kosugi, M.; Kawarabuki, R.; Kawata, M. Investigation for removal of organic carbon from carbonaceous copper sulphide ore and improving the recovery of copper through flotation. In *Characterization of Minerals, Metals, and Materials*; Springer: Cham, Switzerland, 2018; p. 343. [CrossRef]
8. Miller, J.D.; Wan, R.Y.; Diaz, X. Preg-robbing gold ores. In *Gold Ore Processing*; Elsevier: Amsterdam, The Netherlands, 2016; p. 885. [CrossRef]
9. Feng, D.; Van Deventer, J.S.J. Preg-robbing phenomena in the thiosulphate leaching of gold ores. *Miner. Eng.* **2001**, *14*, 1387. [CrossRef]
10. Pan, Z.; Xiong, J.; Cui, Y.; Wei, Q.; Jia, W.; Zhang, Z.; Qin, W. Effect mechanism of CMs on the flotation separation of lead–zinc ore. *Sep. Purif. Technol.* **2022**, *294*, 121101. [CrossRef]
11. Niu, H.; Yang, H.; Tong, L.; Zhong, S.; Liu, Y. Spectral study of humic substance extract from pressurized oxidizing slag of Carlin-typed gold deposit. *J. Phys. Conf. Ser.* **2019**, *1347*, 012027. [CrossRef]
12. Moroz, T.N.; Ponomarchuk, V.A.; Goryainov, S.V.; Palchik, N.A.; Edwards, H.G.; Zhmodik, S.M. Raman spectra of natural CMs from a black shale formation. *J. Raman Spectrosc.* **2015**, *46*, 959. [CrossRef]
13. Heckmann, A.; Fromm, O.; Rodehorst, U.; Münster, P.; Winter, M.; Placke, T. New insights into electrochemical anion intercalation into CMs for dual-ion batteries: Impact of the graphitization degree. *Carbon* **2018**, *131*, 201. [CrossRef]

14. Morcos, I. On contact angle and dispersion energy of the cleavage graphite/water system. *J. Colloid Interface Sci.* **1970**, *34*, 469. [CrossRef]
15. Bhattacharyya, D.; Depci, T.; Assemi, S.; Prisbrey, K.; Miller, J.D. The nature of graphene surfaces as determined from the wettability studies of basal and edge planes. *ECS Trans.* **2015**, *66*, 45. [CrossRef]
16. Sime, F.M.; Jin, J.; Wang, X.; Wick, C.D.; Miller, J.D. Characterization and simulation of graphite edge surfaces for the analysis of CM separation from sulphide ores by flotation. *Miner. Eng.* **2022**, *182*, 107590. [CrossRef]
17. Spieth, V. Zechstein Kupferschiefer at Spremberg and Related Sites: Hot Hydrothermal Origin of the Polymetallic Cu–Ag–Au Deposit. Ph.D. Thesis, Universität Stuttgart, Stuttgart, Germany, 2019. [CrossRef]
18. Smith, T.; Lin, D.; Lacouture, B.; Anderson, G. Removal of organic carbon with a Jameson Cell at Red Dog Mine. In Proceedings of the 40th Annual Canadian Mineral Processors Conference, Ottawa, ON, Canada, 22–24 January 2008; Available online: https://www.glencoretechnology.com/.rest/api/v1/documents/ed2ef91249f47c139df4a66f1f12ec0e/Removal-of-Organic-Carbon-with-a-Jameson-Cell-at-R.pdf (accessed on 17 August 2025).
19. Konieczny, A.; Pawlos, W.; Krzeminska, M.; Kaleta, R.; Kurzydlo, P. Evaluation of organic carbon separation from copper ore by pre-flotation. *Physicochem. Probl. Miner. Process.* **2013**, *49*, 189. [CrossRef]
20. Gredelj, S.; Zanin, M.; Grano, S.R. Selective flotation of carbon in the Pb–Zn carbonaceous sulphide ores of Century Mine, Zinifex. *Miner. Eng.* **2009**, *22*, 279. [CrossRef]
21. Brooke, K.; Bullock, N.; Harvey, R.; O'Sullivan, R.; Phan, C.; Tan, P.; Telford, P.; Edgar, M. Mount Isa and Townsville operations. In *Australasian Mining and Metallurgical Operating Practices*, 3rd ed.; Rankin, W.J., Ed.; AusIMM: Carlton, Australia, 2013; Volume 1, p. 931.
22. Martinez-Esparza, G. Productive Versus Non-Productive Porphyry Systems Surrounding the Peñasquito Diatreme-Porphyry System. Ph.D. Thesis, University of Nevada, Reno, NV, USA, 2020; p. 288. Available online: https://search.proquest.com/openview/203d4bda5c89adb9da0e377d657fb9f4/1?pq-origsite=gscholar&cbl=51922&diss=y (accessed on 17 August 2025).
23. Polat, M.; Polat, H.; Chander, S. Physical and chemical interactions in coal flotation. *Int. J. Miner. Process.* **2003**, *72*, 199–213. [CrossRef]
24. Xu, M.; Li, C.; Wang, Y.; Zhang, H. Investigation on mechanism of intensifying coal fly ash froth flotation by pretreatment of non-ionic surfactant. *Fuel* **2019**, *254*, 115601. [CrossRef]
25. Kadagala, M.R.; Nikkam, S.; Tripathy, S.K. A review on flotation of coal using mixed reagent systems. *Miner. Eng.* **2021**, *173*, 107217. [CrossRef]
26. Mierczynska-Vasilev, A.; Beattie, D.A. Adsorption of tailored carboxymethyl cellulose polymers on talc and chalcopyrite: Correlation between coverage, wettability, and flotation. *Miner. Eng.* **2010**, *23*, 985–993. [CrossRef]
27. Chimonyo, W.; Fletcher, B.; Peng, Y. The differential depression of an oxidised starch on the flotation of chalcopyrite and graphite. *Miner. Eng.* **2020**, *146*, 106114. [CrossRef]
28. Chen, X.; Liu, S.; Peng, Y. A new approach to selectively reject naturally hydrophobic gangue in the flotation of base metal sulphide minerals. *Min. Metall. Explor.* **2021**, *38*, 713–720. [CrossRef]
29. Glembozkii, V.A.; Dmitrieva, G.M.; Sorokin, M.M. *Nonpolar Reagents and Their Effect during Flotation*; Nauka: Moscow, Russia, 1968. (In Russian)
30. *State Standard 2408.3-90*; Solid Fuel. Methods for Determination of Oxygen; Standartinform: Moscow, Russia, 1990; 16p. (In Russian)
31. *State Standard 11022-95*; Solid Mineral Fuels. Methods for Determination of Ash; Standartinform: Moscow, Russia, 2006; 8p. (In Russian)
32. Fuerstenau, M.C.; Clifford, K.L.; Kuhn, M.C. The role of zinc–xanthate precipitation in sphalerite flotation. *Int. J. Miner. Process.* **1974**, *1*, 307–318. [CrossRef]
33. Pugh, R.J. Macromolecular organic depressants in sulphide flotation—A review, 1. Principles, types and applications. *Int. J. Miner. Process.* **1989**, *25*, 101–130. [CrossRef]
34. Sosipatorov, A.I.; Panchenko, G.M.; Vysotin, V.V.; Vinokurova, M.A.; Chikin, A.Y. Application prospects of domestic depressor reagent under carbonaceous gold-bearing ore flotation. *Proc. Irkutsk. State Tech. Univ.* **2018**, *22*, 184–193. (In Russian)
35. Ansari, A.; Pawlik, M. Floatability of chalcopyrite and molybdenite in the presence of lignosulfonates. Part I. Adsorption studies. *Miner. Eng.* **2007**, *20*, 600–608. [CrossRef]
36. Ansari, A.; Pawlik, M. Floatability of chalcopyrite and molybdenite in the presence of lignosulfonates. Part II. Hallimond tube flotation. *Miner. Eng.* **2007**, *20*, 609–616. [CrossRef]
37. Gutiérrez, L.; Uribe, L.; Hernández, V.; Vidal, C.; Mendonça, R.T. Assessment of the use of lignosulfonates to separate chalcopyrite and molybdenite by flotation. *Powder Technol.* **2020**, *359*, 216–225. [CrossRef]
38. Kelebek, S.; Yoruk, S.; Smith, G.W. Wetting behaviour of molybdenite and talc in lignosulphonate/MIBC solutions and their separation by flotation. *Sep. Sci. Technol.* **2001**, *36*, 145–157. [CrossRef]

39. Nechiporenko, A.P. *Acid-Base Properties of the Surface of Solid Substances: Methodical Instructions*; LTI im. Lensoveta: Leningrad, Russia, 1989. (In Russian)
40. Nechiporenko, A.P. Donor-Acceptor Properties of the Surface of Solid-Phase Systems. In *Indicator Method: Tutorial*; Lan: St. Petersburg, Russia, 2017. (In Russian)
41. Atchabarova, A.A.; Abdimomyn, S.K.; Abduakhytova, D.A.; Zhigalenok, Y.R.; Tokpayev, R.R.; Kishibayev, K.K.; Khavaza, T.N.; Kurbatov, A.P.; Zlobina, Y.V.; Djenizian, T.J. Role of carbon material surface functional groups on their interactions with aqueous solutions. *J. Electroanal. Chem.* **2022**, *922*, 116707. [CrossRef]
42. Tanabe, K. *Solid Acids and Bases: Their Catalytic Properties*; Elsevier: Amsterdam, The Netherlands, 2012.
43. Mao, L.; Yoon, R.-H. Predicting flotation rates using a rate equation derived from first principles. *Int. J. Miner. Process.* **1997**, *51*, 171–181. [CrossRef]
44. Kijewski, P.; Leszczyński, R. Węgiel organiczny w rudach miedzi—Znaczenie i problemy. *Zesz. Nauk. Inst. Gospod. Surowcami Miner. Energią PAN* **2010**, *79*, 131–146. (In Polish)
45. Chimonyo, W.; Peng, Y. Review of the Influence of Native and Modified Biopolymers on Carbonaceous Gangue Depression in Selective Flotation. *Miner. Process. Extr. Metall. Rev.* **2025**, *46*, 593–610. [CrossRef]
46. Yergeshev, A.; Tokpayev, R.; Karmeeva, M.; Khavaza, T.; Yergesheva, N.; Atchabarova, A.; Nauryzbayev, M.; Ignatkina, V. Technological Properties Contrast of Galena, Sphalerite, CM and Choice of Flotation Technology. *Minerals* **2025**, *15*, 883. [CrossRef]

minerals

MDPI

Article

The Recycling of Plastics and Current Collector Foils from End-of-Life NMC-LCO Type Electric Vehicle Lithium-Ion Batteries Using Selective Froth Flotation

Fulya Mennik [1], Nazlım İlkyaz Dinç [1], Beril Tanç Kaya [2], Zoran Štirbanović [3], Ronghao Li [4] and Fırat Burat [1,*]

[1] Faculty of Mines, Mineral Processing Engineering Department, Istanbul Technical University, Maslak, Istanbul 34467, Türkiye; mennik17@itu.edu.tr (F.M.); dincnaz@itu.edu.tr (N.İ.D.)
[2] Faculty of Mines, Geological Engineering Department, Istanbul Technical University, Maslak, Istanbul 34467, Türkiye; tancb@itu.edu.tr
[3] Technical Faculty in Bor, Department of Mineral and Recycling Technologies, University of Belgrade, Vojske Jugoslavije 12, 19210 Bor, Serbia; zstirbanovic@tfbor.bg.ac.rs
[4] Department of Mining and Materials Engineering, McGill University, 3610 University Street, Montreal, QC H3A 0C5, Canada; ronghao.li@mail.mcgill.ca
* Correspondence: buratf@itu.edu.tr; Tel.: +90-532-6466658

Abstract

The recent increase in end-of-life (EoL) lithium-ion batteries (LiBs) has become a significant concern worldwide. Most studies in the literature have primarily focused on recovering cathode active metals from black mass (BM), whereas the separation of anode–cathode foils, plastics, and casing metals which are the essential components of LiBs has received relatively little attention. To reduce costs and maximize the recovery of valuable metals in subsequent hydrometallurgical or pyrometallurgical processes, EoL LiBs require appropriate pre-treatment. This study aims to scrape off the BM adhering to the electrode foils resulting from gradual crushing and subsequently separate the plastics and copper (Cu) from other metals through a two-step selective flotation process. The results demonstrated that plastics, due to their natural hydrophobicity, could be effectively removed using a frother, achieving more than 95% recovery with less than 5% metallic contamination. Following plastic flotation, Cu particles were floated in the presence of 3418A, yielding a Cu concentrate containing 65.13% Cu with a recovery rate of 96.4%. Additionally, the aluminum (Al) content in the non-floating material, remaining in the cell, increased to approximately 77%.

Keywords: lithium-ion battery; plastics; copper; aluminum; flotation; recycling

1. Introduction

Developments in energy storage technologies have brought significant transformations across various sectors and have become crucial for achieving a sustainable future. Moreover, the rising energy demand, increased use of renewable energy sources, and extensive adoption of electric vehicle (EV) technologies have contributed to the need for advanced energy storage systems. In particular, high fossil fuel consumption and the necessity to reduce environmental damage have driven efforts to develop renewable and sustainable energy sources such as high-quality batteries, wind turbines, fuel cells, and solar panels [1–3].

LiBs are characterized by their compact size, lightweight design, high energy density and voltage, extended lifespan, absence of memory effect, efficient operation across a wide temperature range, and eco-friendly properties [1,4]. The growing use of EVs and consumer electronics has significantly increased the demand for LiBs, resulting in a substantial rise in the volume of EoL batteries requiring disposal [5]. For example, China produced approximately 15 billion LiBs in 2014, and by 2024, its annual production capacity surpassed 1170 GWh, representing around 76% of global output [6]. This surge in LiB demand highlights their critical role in energy storage, EV charging infrastructure, and large-scale battery energy storage systems (BESS). Consequently, the projected increase in LiB production is expected to lead to a significant volume of EoL batteries.

Incorrect disposal of spent LiBs poses severe environmental problems, including soil and water contamination from leakage of heavy metals and hazardous compounds. Beyond environmental and economic benefits, exploiting the resource potential of battery scraps is essential for conserving natural resources and promoting the sustainable development of metals and related industries [7,8]. Additionally, EoL LiBs contain many components with high economic value, such as lithium (Li), cobalt (Co), copper (Cu), aluminum (Al), nickel (Ni), iron (Fe), manganese (Mn), and graphite [9,10].

LiBs are manufactured to meet specific application requirements, with the most variation occurring in the active cathode material. Several types of LiBs have been developed, including lithium cobalt oxide (LCO), nickel cobalt manganese oxide (NMC), nickel cobalt aluminum oxide (NCA), lithium manganese oxide (LMO), lithium iron phosphate (LFP), and lithium titanate oxide (LTO). Among these, NMC, LCO, LFP, and LMO batteries are the most widely used in commercial applications [11]. Figure 1 illustrates the main components of a typical LiB cell, including the cathode and anode coatings (BM), current collector foils, separator, electrolyte, and casing. NMC-based LiBs generally comprise 5%–20% Co, 5%–7% Li, 7%–10% Cu, 10%–15% Al, 4%–5% Ni, 4%–5% Mn, around 15% organic compounds, and approximately 7% plastic materials, indicating considerable potential for resource recovery [12,13]. Therefore, EoL LiBs are increasingly regarded as strategic secondary raw materials. Recycling these batteries enables the reclamation of critical metals, thus promoting sustainable resource management and contributing to the circular economy [14,15].

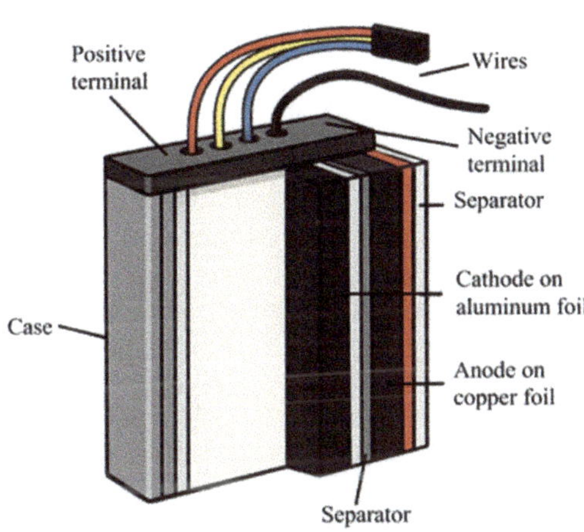

Figure 1. Schematic illustration of the main components of a typical Li-ion battery cell.

Driven by technological advancements, increasing market demand, and the growing volume of waste batteries, global battery recycling capacity has expanded from 9 GWh in 2019 to 50 GWh in 2024, and it is projected to exceed 230 GWh by 2030 [16]. Recycling LiBs facilitates the recovery of critical metals, reducing dependency on mining, conserving natural resources, and supporting sustainable production models. It is estimated that recycling 500,000 tons of LiBs could recover approximately 45,000 tons of Cu, 15,000 tons of Al, 90,000 tons of Fe, 75,000 tons of Li, 60,000 tons of Co, and 35,000 tons of plastic materials [17].

The recycling process of LiBs typically begins with a series of pre-treatment steps designed to prepare dismantled EoL battery packs for main recovery. These steps are crucial for purifying core components such as cathode and anode materials, enhancing the efficiency and safety of subsequent processing [18]. The primary aim of pre-treatment is to achieve effective and safe separation of battery components, which is particularly important for hydrometallurgical recovery and direct recycling methods [19]. Due to the structural complexity and hazardous nature of LiBs, pre-treatment is essential to mitigate safety risks and facilitate the selective recovery of valuable materials. Standard pre-treatment operations include deep discharging to eliminate residual charge, mechanical size reduction, particle classification, physical component separation, thermal and chemical treatments, and cathode material isolation [19–21]. This stage establishes foundational conditions to ensure high recovery yields and environmental compliance in advanced recycling technologies.

Size reduction is crucial for effective liberation of electrode materials and directly influences the efficiency of subsequent purification and enrichment stages. Shredders, hammer mills, and impact crushers are commonly used to break down battery cells into smaller fragments [8,14,22,23]. Further size reduction, if needed, can be performed by grinding into a fine powder. Screening systems then classify materials by particle size, separating fine fractions, rich in Li, Co, and graphite, from coarser fractions containing plastics, Fe, Cu, and Al. The fine fraction (undersize) comprises valuable BM suitable for hydrometallurgical or direct recovery, while the coarse fraction (oversize) can be further processed or recycled, depending on its composition.

The primary recycling technologies for LiBs include pyrometallurgical, hydrometallurgical, and direct recovery methods [24,25]. Pyrometallurgical processes operate at high temperatures, resulting in high energy consumption and the emission of harmful gases [26]. Hydrometallurgical methods are favored for lower energy use, higher recovery efficiency, and reduced environmental impact [27]. These acid leaching processes typically use sulfuric acid, with recent research focusing on replacing traditional reducing agents like hydrogen peroxide with environmentally friendly organic compounds such as glucose, tea waste, and grape seed extracts [2]. Despite their benefits, these methods often involve complex chemical steps, high energy demands, or limited recovery yields.

Flotation separates materials based on their hydrophobic (water-repelling) and hydrophilic (water-attracting) properties by causing some particles to float while others sink. Flotation of coarse metallic and plastic fractions from EoL LiBs shares many mechanistic similarities with classical ore beneficiation. Recent work by Aarab et al. [28] on the eco-friendly flotation of apatite using bio-based collectors demonstrates the potential of sustainable reagent design and adsorption mechanisms that are also relevant to plastic–metal separation. Originally used for ore beneficiation, flotation has recently been applied effectively for the direct recycling of EoL LiBs [29]. This method is advantageous due to its low energy consumption, minimal chemical requirements, and ability to separate graphite from cathode metal oxides based on surface characteristics. The naturally hydrophobic graphite anode facilitates the separation of carbon-based materials in BM [30]. In flotation, graphite particles attach to bubbles and float, while mostly hydrophilic cathode materials like LCO and NMC sink to the bottom fraction [31].

Despite growing interest in LiB recycling, studies focusing on the selective recovery of coarse metallic fractions such as Cu, Al, Fe, and Ni are scarce. Saneie et al. [32] demonstrated effective separation of Cu and Al foils in the −4 + 1 mm size range from 18,650-type spent LiBs using a two-stage flotation process involving initial plastic removal followed by Cu flotation. To our knowledge, only a limited number of studies have reported flotation-based separations relevant to spent LiBs; most target electrode powders/graphite or small-device batteries rather than EV-grade mixed NMC/LCO coarse fractions. Prior research on flotation-based recovery from waste printed circuit boards (PCBs) [33–35] suggests that liberated plastic particles can be effectively separated from each other and metallic components in coarse fractions using froth flotation, indicating promising potential for LiB recycling.

Accordingly, this study focuses on the selective separation of plastics, Cu, Al, and structural metals from NMC and LCO-type LiBs through the application of various surface-active reagents. After initial BM separation from the metallic fraction via sequential size reduction and sieving across four stages, the first flotation step targeted the selective removal of naturally hydrophobic plastics using a frother (Methyl Isobutyl Carbinol, MIBC). In the subsequent flotation stage, the effect of different collectors on Cu recovery was examined by promoting hydrophobic Cu particles to the froth phase, allowing hydrophilic Al and structural metal particles to remain in the pulp.

2. Experimental Studies

2.1. Materials

In this study, approximately 60 kg of mixed NMC and LCO-type EoL EV LiBs were supplied by Exitcom Co., located in Kocaeli, Türkiye. A multistage size reduction and classification system was applied, as illustrated in Figure 2. Initially, the batteries were fully discharged for 48 h using a 5% NaCl solution, and then air-dried. Following this, primary crushing was performed using a dual-shaft shredder at the recycling company to ensure safety. The crushed material (particles smaller than 40 mm) was then transferred to the Mineral Processing Engineering Department's Prof. Dr. Güven Önal Pilot Plant at Istanbul Technical University for subsequent comminution processes.

The BM 1 sample was initially separated using a 0.2 mm sieve because this cut size efficiently removed most of the liberated BM while retaining metallic foils and plastics for subsequent flotation. This choice was based on preliminary screening tests and is independent of the crusher settings. The coarse fraction was subsequently reduced to particles smaller than 10 mm using a shredder. The crushed material was screened again with a 0.2 mm sieve, and the oversized fraction was fed into a four-blade cutting mill (RAM200 model, supplied by RANTEK Co., Ankara, Türkiye) to further scrape off the residual BM adhering to the current collectors. This cutting mill is designed to uniformly crush elastic, soft, medium-hard, fibrous, and heterogeneous materials. The shear forces reduce particle size while maintaining the flat surface shape of the material, which is advantageous for the flotation process. Particle size adjustment was achieved by using interchangeable sieves (6, 4, 2 mm, etc.) located in the lower compartment of the mill. The BM adhering to the surfaces of the current collectors was further collected using a 0.2 mm sieve and preserved for future studies.

The particle size range of −2 + 0.2 mm was selected based on preliminary sieving tests and chemical analysis, which indicated that this fraction contained sufficiently liberated plastic and metal particles for efficient separation. The finer fraction (<0.2 mm) contained most of the BM powders and electrode coatings, which are more suitable for chemical leaching rather than flotation. Although flotation is conventionally applied to particles below ~0.5 mm due to buoyancy limitations of larger spherical particles, the plate-like shape of electrode foils and plastic pieces in this waste material enabled successful flotation at coarser sizes. Additional tests conducted with the −4 + 0.2 mm fraction resulted in poor separation efficiency due to excessive residual active material and insufficient liberation, confirming −2 + 0.2 mm as the optimal size range for flotation.

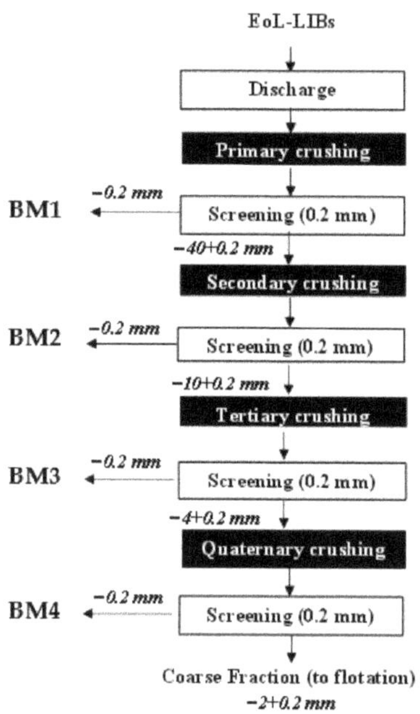

Figure 2. The graph of the multi-stage crushing process.

The coarse fraction, constituting approximately 32% of the feed material, mainly consisted of Cu foils, Al foils, battery casings, plastics, and separators, and was used in characterization and flotation experiments. KAX (Potassium Amyl Xanthate, >98%) (ECS chemicals, Türkiye) and Aerophine-3418A, Aero-3739, Aerofloat-242, Aerofloat-211 (Syensqo, Türkiye), and MIBC (ECS chemicals, Türkiye) were employed as collectors and frother, respectively, in the flotation tests. All flotation tests were performed using municipal tap water from the Terkos system (Istanbul), with a conductivity of approximately 620 μS/cm and pH around 7.5. The main dissolved species were Ca^{2+} (≈85 mg/L), Mg^{2+} (≈12 mg/L), Na^{+} (≈28 mg/L), Cl^{-} (≈30 mg/L), and HCO_3^{-} (≈230 mg/L). The natural pulp pH during flotation was 8.3 ± 0.1 and remained stable throughout the experiments. Batteries were safely discharged in 5 wt.% NaCl solution according to widely accepted recycling practice. Although residual chlorides could potentially adsorb onto current collector surfaces, the subsequent comminution and water-based slurry preparation steps are expected to remove most soluble salts before flotation.

After quartering the comminuted material, a representative sample was collected, and sieve analysis was conducted using an electric shaker with standard sieves. The experimental data were interpreted based on product content and recovery. The compositions of major and trace elements were studied via a BRUKER S8 TIGER model wavelength dispersive X-ray fluorescence spectrometer (WDXRF) at the Geochemistry Research Laboratory of the ITU Department of Geological Engineering. For this research, milled samples were homogenized with wax at a 5:1 (*w*/*w*) ratio and pelletized using a HERZOG model pelletizer to create homogeneous discs suitable for XRF measurements. The accuracy of the analytical method was evaluated by analyzing the certified reference material, CRM (USGS DNC-1A and GEOSTATS GBM915-4), under the same sample preparation and measurement conditions. Three replicate analyses were performed, and the mean values were compared with the certified and/or reference values; the measured values showed good agreement with the certificate. The detection limits are 100 ppm for Al, Mn, and Fe, and 4.7, 2.2, and 1.5 ppm for Cu, Co, and Ni, respectively. The Loss on Ignition (LOI) study was performed in two consecutive stages to ascertain the volatile content of the samples. Initially, 2–3 g of material was combusted at 500 °C for 2 h to quantify organic matter and polymeric binders, including Polyvinylidene Fluoride (PVDF) and separator residues. Thereafter, the identical samples were combusted at 1000 °C for an additional 2 h to evaluate the total volatile content, encompassing carbonates and other thermally unstable constituents. This two-step LOI methodology facilitated a more thorough assessment of both organic and inorganic volatiles, especially in the BM and flotation-derived products. This approach is commonly reported for battery recycling studies to quantify Li_2CO_3 and other carbonates, which release CO_2 at temperatures > 800 °C. Although some volatilization of fluorides at 1000 °C is possible, we minimized exposure time to limit such effects.

XRF is constrained in its ability to detect light elements (Z < 11), including carbon (C), Li, oxygen (O), and fluorine (F), owing to the low-energy characteristics of their X-rays and matrix absorption phenomena [35,36]. Consequently, other methodologies were employed: lithium was measured via inductively coupled plasma (ICP) analysis using a PerkinElmer Avio 200 instrument, whereas C, O, and F were aggregated under the "others" group.

The recovery (*R*) used in result interpretation was calculated by Equation (1):

$$R\ (\%) = (Cc/Ff) \times 100 \quad (1)$$

where ***C*** is the weight of the concentrate, ***c*** is the metal content of the concentrate, ***F*** is the weight of the feed, and ***f*** is the metal content in the feed. All flotation tests were performed once under identical operating conditions, while the chemical assays for each product stream were conducted in triplicate to ensure analytical accuracy. In addition, the flotation experiment that yielded the best Cu recovery result (using 3418A) was repeated three times under the same conditions, and the results were consistent within experimental error.

To investigate the collector adsorption on Cu and Al electrode foils, a scanning electron microscopy (SEM) (Hitachi SU3500 model, Japan) coupled with an 80 mm^2 XMaxN silicon drift detector energy-dispersive spectroscopy (EDS) system (Oxford Instruments, Oxfordshire, UK). Cu collector electrode foils, whose surfaces were mostly cleaned of graphite by brushing from the anode side, and Al collector electrodes with cathode material still attached were used in the SEM studies. Sample preparation ensured that, after comminution and classification, the Cu foil surfaces were largely free of graphite due to the weak binding agent. In contrast, the cathode material is strongly adhered to the Al foil by the PVDF binder, making mechanical removal potentially damaging; therefore, Al samples were used without scraping. The foils were soaked for 10 min in 3000 g/t 3418A solution. After careful drying, they were cut into 5 × 5 mm pieces and mounted on aluminum stubs using double-sided carbon tape for SEM analysis. EDS results were interpreted qualitatively, as the quantification of light elements is known to be unreliable and susceptible to contamination. Therefore, the observed increases in surface C and S after collector conditioning are considered indicative of relative trends rather than absolute values. Both secondary electron (SE) and backscattered electron (BSE) signals were collected at an accelerating voltage of 20 kV under high vacuum. During BSE-EDS analysis, the detector deadtime was maintained at approximately 25%, and count rates exceeded 40,000 counts per second (cps). The collected data were analyzed using Aztec software (Version 3.0, Oxford Instruments, UK) to identify elements present in the samples.

2.2. Beneficiation Methods

Before the flotation experiments, 50 g of the sample was weighed and added to 1000 mL of tap water, corresponding to a pulp density of approximately 5 wt.% solids, then mixed at 1500 rpm for 15 min to remove any remaining BM. The relatively low pulp density was deliberately selected because, compared with conventional ores where higher pulp densities are common, this material contains a significant proportion of plastic particles and fine BM, which would otherwise increase slurry viscosity and promote unwanted entrainment at higher solids. Operating at lower pulp density results in a cleaner froth product and improves the selectivity of plastic flotation in the first stage.

The dispersed BM was separated using a 74 μm sieve, and the oversize material was fed into a 1.5 L self-aerated Denver-type flotation machine operating at an impeller speed of 1200 rpm. No pH modifiers were used to maintain a more environmentally friendly process. City tap water was utilized throughout the experiments, with the natural pulp pH measured at approximately 8.3.

The choice of collectors was guided by their well-documented performance in sulfide and base-metal flotation systems. KAX was included as a reference sulfhydryl collector to benchmark Cu flotation response. Aerophine-3418A, a phosphine-based collector, was selected for its known selectivity towards Cu minerals and lower affinity for iron and aluminum, which was expected to enhance Cu foil recovery with minimal contamination. Aerofloat-211 and Aerofloat-242 (dithiophosphates) were tested to evaluate whether stronger adsorption on Cu surfaces could further improve recovery or concentrate grade. Compared structurally to xanthates, but using phosphorus (P) instead of C, making a P-based sulfur (S) collector with improved selectivity. Sodium di-isobutyldithiophosphinate-based collector is primarily composed of sodium (Na), P, S, C, and hydrogen (H). Its approximate molecular formula is $C_8H_{18}NaPS_2$ [37,38]. Importantly, all collectors chosen are commercial reagents widely used in large-scale natural Cu ore processing, with well-understood chemistry and minimized environmental drawbacks, making the results more readily transferable to industrial LiB recycling operations.

Once the reagents had sufficient contact with the particles, the air valve was opened to introduce an airflow of 3 L/min, and the floated particles were collected during a 3 min flotation period. The first-stage flotation aimed to separate the naturally hydrophobic plastic fraction from the metallic components. MIBC was added in five equal charges of 100 g/t for each stage (5 mg per step for 50 g feed) 60 s before aeration, mixing at 1200 rpm (room temperature, ~22 ± 2 °C). Following plastic separation, Cu particles remaining in the sink fraction were floated using various collector reagents (KAX, 3418A, Aero-3739, Aerofloat-242, and Aerofloat-211). During Cu flotation, collectors were dosed totally at 3000 g/t (three equal charges of 1000 g/t for each stage, 50 mg per addition, 150 mg total per test) with 5 min conditioning at 1200 rpm after each addition. No additional cleaning or scavenger stages were applied to better assess the intrinsic selectivity of the flotation reagents under rougher-only conditions.

3. Results

3.1. Characterization Studies

Elemental analysis of the LiB sample used in the experimental studies revealed the presence of several key metals. As summarized in Table 1, the concentrations of Cu, Al, Co, Ni, and Mn in the initial feed material were 10.38%, 15.78%, 26.94%, 4.23%, and 4.01%, respectively. These metals originate primarily from current collectors (Cu, Al) and active electrode materials (Co, Ni, Mn). Following staged comminution and particle size classification, a significant redistribution of these metals was observed across different size fractions. Cu and Al, which serve as the primary current collector foils, became increasingly concentrated in the coarse fraction (−2 + 0.2 mm), representing approximately 25% of the feed mass. In this fraction, the combined Cu and Al content reached nearly 65 wt.%, demonstrating efficient liberation and size-based separation of the foil materials during mechanical processing.

Table 1. Metal contents and recoveries of the products after comminution and classification.

Products	Amount, wt.%	Cu		Al		Fe		Ni	
		C [1], %	D [2], %	C, %	D, %	C, %	D, %	C, %	D, %
−2 + 0.2 mm	31.9	25.58	78.6	39.35	79.5	0.35	54.6	1.73	13.1
BM 1	12.3	2.34	2.8	2.80	2.2	0.10	6.0	3.06	8.9
BM 2	17.6	1.70	2.9	4.10	4.6	0.10	8.6	4.45	18.4
BM 3	19.0	4.81	8.8	4.67	5.6	0.20	18.6	5.87	26.4
BM 4	19.2	3.74	6.9	6.63	8.1	0.13	12.2	7.30	33.2
Total	100.0	10.38	100.0	15.78	100.0	0.20	100.0	4.23	100.0
Products	Amount, wt.%	Co		Li		Mn		Others	
		C, %	D, %	C, %	D, %	C, %	D, %	C, %	D, %
−2 + 0.2 mm	31.9	6.26	7.4	1.54	18.8	1.56	12.4	20.96	19.8
BM 1	12.3	20.83	9.5	2.18	10.3	3.03	9.3	64.07	23.5
BM 2	17.6	36.30	23.6	3.54	23.7	4.02	17.6	44.37	23.1
BM 3	19.0	40.85	28.9	3.43	24.9	5.33	25.3	31.98	18.0
BM 4	19.2	42.90	30.6	3.05	22.3	7.37	35.4	27.40	15.6
Total	100.0	26.94	100.0	2.62	100.0	4.01	100.0	33.72	100.0

[1] Content, [2] Distribution.

By contrast, the fine fraction (−0.2 mm), referred to as "BM", was significantly enriched in Co, Ni, and Mn, metals associated with cathode materials. The Co content increased steadily in finer BM fractions as size reduction progressed, rising from 20.83% in BM 1 to 42.90% in BM 4. This marked enrichment indicates progressive liberation and concentration of Co-bearing cathode materials during successive comminution and classification steps. This process likely reflects the effective detachment of Co from Al current collector foils, governed by the breakdown of strong polymeric binders (e.g., PVDF) at later milling stages.

The carbonaceous content (graphite), primarily coated onto Cu foil, was more readily detached during initial crushing steps due to its relatively weak interfacial bonding. This is evidenced by a high "others" fraction (mainly graphite, plastic, and residual binder) of up to 64.07 wt.% in BM 1. Moreover, it was observed that intensive comminution not only liberated the active materials but also fragmented the Cu and Al foils into fine particles, thereby contaminating the BM. This contamination was particularly pronounced in BM 3 and BM 4, where the intrusion of Cu and Al into the fine fractions increased significantly. These findings indicate that the multi-stage size reduction process causes foil-structured Cu and Al metals to migrate into the fine fractions, leading to contamination of the BM and potentially reducing overall metal recovery efficiency. A particle size distribution analysis was conducted on the oversize fraction, comprising mainly Cu, Al, plastic materials, and casing metals, retained on the sieve after comminution and classification. The particle size characteristics of this material, which was subsequently used in flotation experiments, were quantified by determining the d_{80} and d_{50} values, measured at 0.95 mm and 0.62 mm, respectively. These size parameters are critical for optimizing flotation performance, as they affect the hydrodynamic behavior and surface interactions of particles during the separation process.

As illustrated in Figure 3, the highest concentrations of Cu and Al were detected in the −0.8 + 0.5 mm size fraction. An increasing trend in the content of cathode active metals was observed toward the finer fractions. This suggests that finer Al foil particles retain a considerable amount of unreleased cathode material, indicating incomplete liberation during the comminution process.

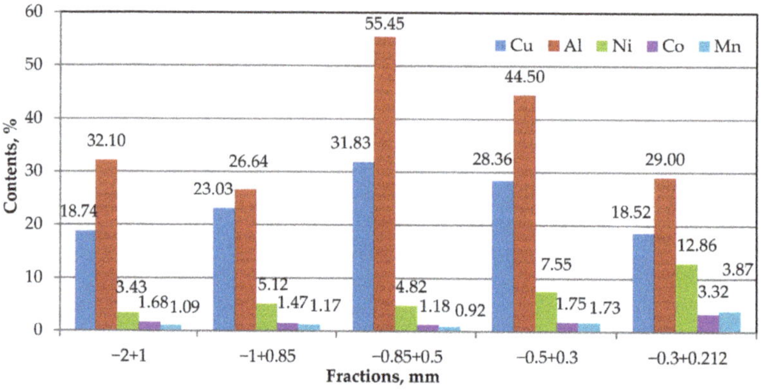

Figure 3. Schematic illustration of the flotation stages.

SEM analysis revealed notable changes in surface morphology and composition of the foils after conditioning with 3418A. As shown in the SE images in Figure 4, the surface roughness of Cu foil increased remarkably after conditioning, suggesting the adsorption of 3418A collector. This could be supported by BSE-EDS results: as shown in Figures 5 and 6, after conditioning, the surface C and S contents increased remarkably, from 7.6% to 26.3% and from non-detectable to 16.4%, respectively. Meanwhile, the Cu content decreased from 91.7% to 24.3%. Surface modifications on Cu foil are primarily due to the adsorption of di-isobutyldithiophosphinate species from 3418A, where the phosphorus–sulfur functional groups form strong chemisorbed complexes on the mineral surface, altering its flotation behavior. In comparison, no significant changes in the surface roughness and composition could be observed on Al foil after conditioning (as shown in Figures 4–6), suggesting the collector did not adsorb onto its surface.

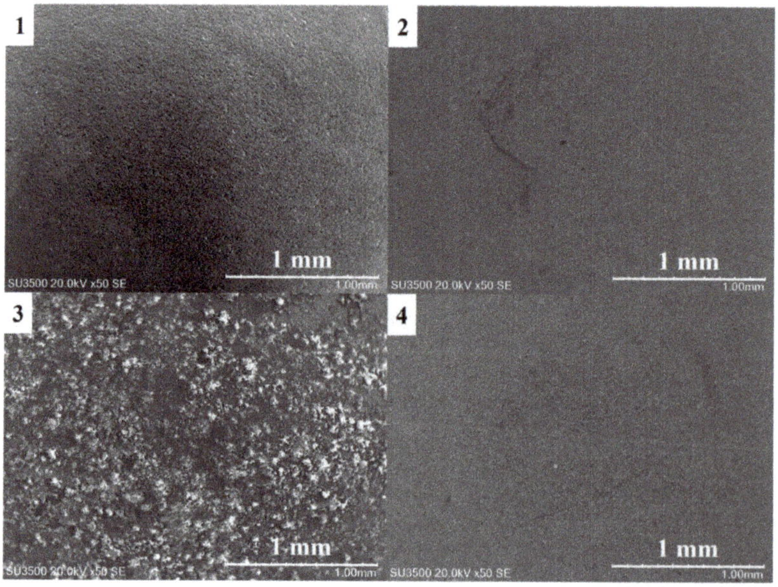

Figure 4. SE images of foils before (#1 Cu and #2 Al) and after (#3 Cu and #4 Al) conditioning.

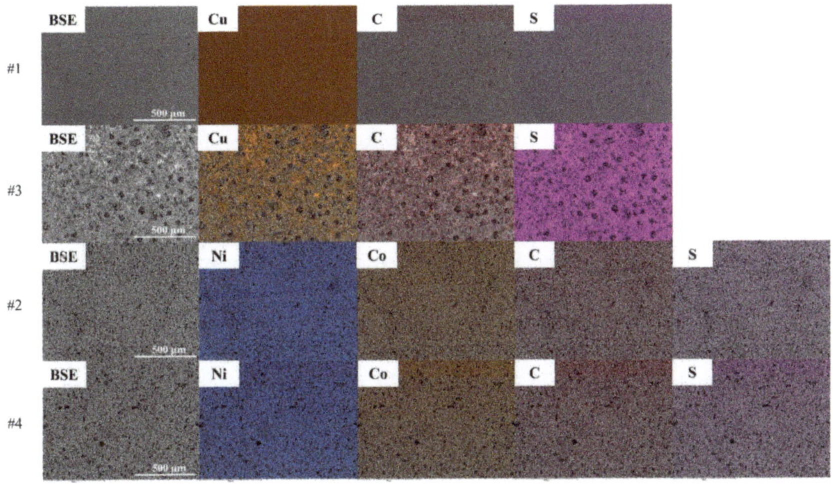

Figure 5. BSE images of major elements present in foils before (#1 Cu and #2 Al) and after (#3 Cu and #4 Al) conditioning.

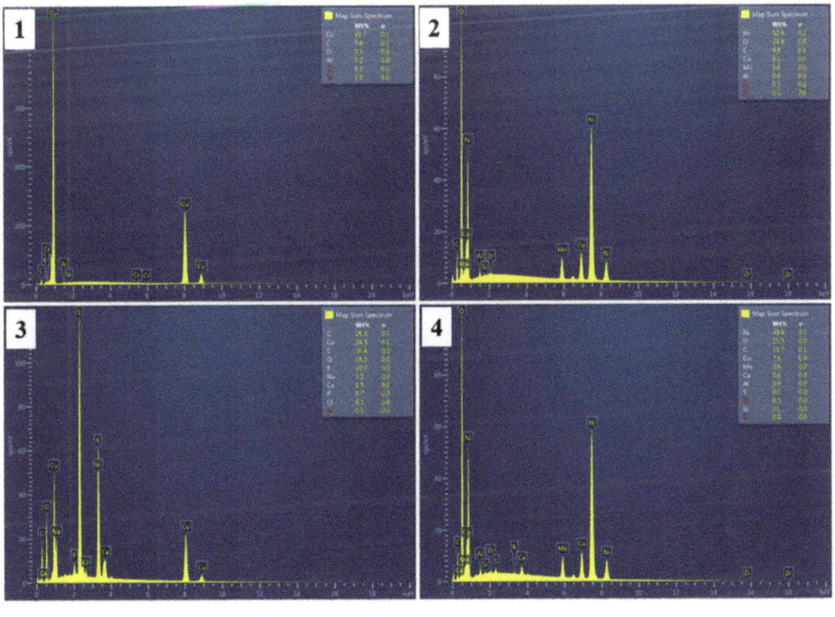

Figure 6. BSE-EDS semi-quantitative results of foils before (#1 Cu and #2 Al) and after (#3 Cu and #4 Al) conditioning with 3418A.

3.2. Flotation Experiments

3.2.1. Plastic Flotation

Plastics are naturally hydrophobic and possess a much lower specific gravity compared to metals. Fadel et al. [39] reported that pure PET exhibits a contact angle of 78.9°, confirming its hydrophobic nature. In froth flotation, selective separation is achieved based on differences in particle hydrophobicity. In our previous study on the selective recovery of metals from spent mobile phone LiBs [40], contact angle measurements were used to assess the surface hydrophobicity of key metallic constituents such as Cu and Al. The relatively low contact angles of unconditioned metallic foils, 50.5° for Cu and 56.1° for Al, support the effectiveness of plastic flotation, where a clear separation between plastics and metals can be realized. Plastic flotation tests were conducted in five stages: no frother was used in the first stage, followed by four stages with 100 g/t MIBC added each time. Without any frother, about 12% of the total plastics floated, increasing to approximately 79% flotation recovery with 200 g/t of MIBC. The cumulative weight of floated plastics systematically increased, reaching 95% at 400 g/t. Due to size reduction processes dominated by shear forces, metal and plastic components adopted plate-like morphologies and became largely liberated from one another. This morphological transformation provides a significant advantage during flotation, facilitating effective separation of plastic particles, which exhibit substantially higher hydrophobicity than metals, even at relatively coarse particle sizes. As shown in Table 2, approximately 16% of the feed material to the first flotation stage was recovered as a plastic product. Approximately 5% by weight of fine-sized Al particles, which exhibit greater floatability than Cu, were transferred to the floated product, whereas Cu particles, with lower floatability than Al, accounted for around 1.5%. A small portion of cathode active materials, still attached to Al foils reported to the plastic fraction.

Table 2. Metal content and recoveries of flotation.

Collector Type	Products	Amount wt.%	Cu		Al		Co		Ni		Mn		Others	
			C [1], %	R [2], %	C, %	R, %	C, %	R, %	C, %	R, %	C, %	R, %	C, %	R, %
KAX	Plastics	14.3	1.25	0.7	12.70	4.7	4.58	10.4	1.30	10.7	1.28	11.8	74.29	48.8
	BM	8.5	1.24	0.4	6.30	1.4	20.18	27.3	5.66	27.6	5.14	28.1	50.96	19.9
	Cu	38.8	62.47	95.8	22.25	22.3	2.25	14.0	1.02	22.8	0.95	23.8	8.53	15.2
	Al	38.5	2.02	3.1	71.84	71.6	7.84	48.3	1.75	38.9	1.46	36.3	9.03	16.0
	Total	100.0	25.28	100.0	38.63	100.0	6.25	100.0	1.73	100.0	1.55	100.0	21.69	100.0
3418A	Plastics	15.7	1.41	0.9	12.68	5.1	4.96	12.5	1.58	13.9	1.46	14.5	73.79	54.0
	BM	10.0	1.33	0.5	6.28	1.6	19.96	32.0	5.51	30.7	5.41	34.2	51.81	24.1
	Cu	37.9	65.13	96.4	21.12	20.7	2.03	12.4	0.73	15.5	0.67	16.1	7.06	12.5
	Al	36.5	1.52	2.2	76.98	72.6	7.33	43.1	1.95	39.9	1.52	35.2	5.55	9.4
	Total	100.0	25.58	100.0	38.70	100.0	6.21	100.0	1.78	100.0	1.58	100.0	21.43	100.0
AERO 3739	Plastics	15.5	1.52	0.9	12.47	4.9	5.03	12.7	1.56	14.1	1.40	14.5	74.03	55.3
	BM	10.1	1.28	0.5	6.20	1.6	19.95	32.8	5.29	31.1	5.35	36.0	52.35	25.4
	Cu	5.9	63.27	15.0	21.97	3.2	1.83	1.8	0.74	2.5	0.63	2.5	8.69	2.5
	Al	68.5	30.25	83.6	52.25	90.3	4.71	52.7	1.31	52.3	1.03	47.1	5.11	16.9
	Total	100.0	24.81	100.0	39.66	100.0	6.13	100.0	1.72	100.0	1.50	100.0	20.77	100.0
AERO 242	Plastics	15.1	1.65	1.0	12.84	4.9	4.70	11.5	1.47	13.4	1.36	13.9	74.00	53.6
	BM	8.0	1.18	0.4	5.57	1.1	19.79	25.5	4.90	23.6	5.55	29.9	52.99	20.3
	Cu	21.4	65.85	56.6	15.93	8.5	1.56	5.4	0.94	12.2	0.78	11.3	12.33	12.7
	Al	55.5	18.88	42.0	61.60	85.5	6.43	57.6	1.52	50.8	1.20	44.9	5.06	13.5
	Total	100.0	24.94	100.0	39.96	100.0	6.19	100.0	1.66	100.0	1.48	100.0	20.86	100.0
AERO 211	Plastics	15.7	1.58	1.0	12.32	4.9	4.83	12.1	1.48	13.2	1.44	14.5	74.12	53.8
	BM	9.9	1.28	0.5	5.48	1.4	19.42	30.7	5.32	29.9	5.35	34.2	52.97	24.4
	Cu	28.6	69.22	80.1	16.21	11.8	1.82	8.3	0.68	11.0	0.60	11.1	8.82	11.7
	Al	45.9	9.91	18.4	69.79	81.8	6.68	48.9	1.76	45.9	1.36	40.2	4.75	10.1
	Total	100.0	24.70	100.0	39.11	100.0	6.26	100.0	1.76	100.0	1.55	100.0	21.55	100.0

[1] Content, [2] Recovery.

Particle surface wettability, whether hydrophilic or hydrophobic, can be effectively changed through surface chemical modifications [40]. Although Al particle surfaces are initially more hydrophobic than Cu surfaces before reagent addition, the hydrophobicity of Cu increases significantly (~95°) and exceeds that of Al (~78°) upon addition of collectors originally designed for natural Cu ore flotation [40]. In EoL LiBs, Cu and Al are typically present as thin, flat monolayer foils (see Figure 7). This unique morphology enables efficient separation through flotation techniques at significantly coarser particle sizes compared to those typically employed in the flotation of primary ores. Subsequently, the residual material within the cell, following plastic flotation, was processed through the Cu flotation circuit by introducing an appropriate dosage of collector and frother to facilitate the selective recovery of Cu-bearing components.

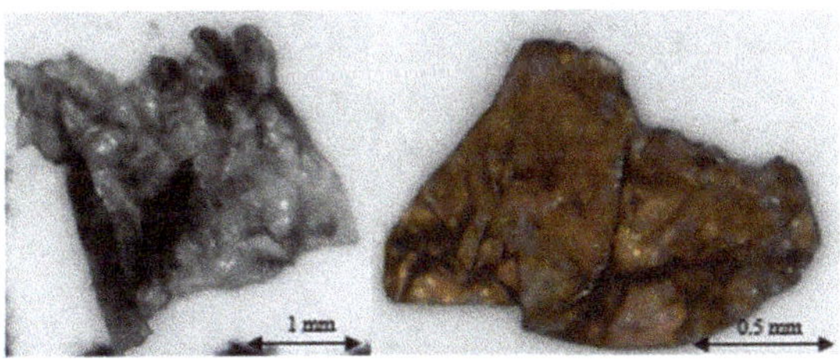

Figure 7. Images of floating products, plastic (**left**) and Cu (**right**), in the first and second flotation stages.

3.2.2. Copper Flotation

The sink fraction from the plastic flotation stage was subjected to Cu flotation using various collectors. 3418A exhibited the highest selectivity and recovery for Cu foils under the applied conditions. The Cu concentrate obtained with 3418A contained 65.13 ± 0.64 wt.% Cu with a recovery of 96.4 ± 0.10%, based on the average of three replicate chemical assays. These results confirm the high grade and recovery previously reported and demonstrate that the data are reproducible within a narrow confidence range.

The phosphorus–sulfur functional groups of 3418A are known to form strong Cu–S and Cu–P bonds, creating a hydrophobic layer that enhances bubble–particle attachment during flotation. The improved Cu floatability observed with 3418A can be attributed to its selective adsorption on Cu foil surfaces. SEM and BSE-EDS analyses (Figures 4–6) revealed increased surface roughness and a notable rise in surface sulfur and carbon content after conditioning, suggesting adsorption of di-isobutyldithiophosphinate species. While these results provide qualitative evidence of surface modification, XPS or FTIR spectroscopy would be more suitable to confirm the chemisorbed collector species and clarify the binding mechanism at the molecular level. This finding is consistent with previous reports on the use of Aerophine collectors in complex sulfide ore systems [37,38]. In contrast, no notable adsorption or surface modification was detected on Al foils, confirming the selectivity of 3418A toward Cu. Furthermore, since the plastic components (PP/PE) are inherently hydrophobic with high contact angles (>75° [39]), they were floated in the first stage without collector addition, and any potential interaction with 3418A in the second stage was minimal. This selective adsorption behavior explains the high Cu grade and recovery obtained, as well as the efficient separation between Cu and Al fractions.

Another commercial reagent commonly preferred in the flotation of natural Cu ores, KAX, also yielded successful results, producing a Cu concentrate assaying 62.47% Cu with a 95.8% Cu recovery. Although the other three collectors produced concentrates with similar Cu grades, their recovery rates were lower due to the reduced quantity of floated material.

When evaluated based on Al content, the best performance was again achieved with the 3418A collector. A sink product containing approximately 77% Al was obtained with a 72.6% Al recovery. The relatively lower Al recovery compared to Cu is attributed to the loss of small-sized Al particles entrained in the Cu concentrate.

Graphite is not typically used in Al current collectors; however, it is predominantly employed as an anode material on Cu surfaces. Considering that some graphite remains adhered to Cu particles after comminution and screening, the "others" category for Cu and

Al in Table 2 likely includes polymer-based separator fragments that were not effectively removed during the initial plastic flotation stage. With the addition of a collector in the Cu flotation circuit, a portion of these particles was recovered, whereas more fibrous and irregularly shaped separator fragments remained in the sink product, often attached to metallic particles. As a potential solution, these insulating plastic materials can be removed either by feeding the material into electrostatic or eddy current separators after drying or by subjecting them to a scavenger flotation circuit in the presence of suitable collectors and frothers.

3.3. Mass Balance and Overall Flowsheet

Based on the characterization and enrichment results, a comprehensive flowsheet was developed for recovering valuable metals from EoL LiBs sourced from EVs (Figure 8). Following an initial discharging stage, the EoL LiB sample underwent a four-stage comminution process. The BM, constituting 68.1% of the total weight, was then separated via sieving. The −2 + 0.2 mm particle size fraction was subjected to froth flotation, during which plastic components were selectively recovered using 500 g/t MIBC as a frother. The floated plastic fraction accounted for 5.0% of the feed and contained negligible metal content, making it suitable for disposal or potential secondary use in other industrial applications (e.g., energy recovery, composite filler production), thus reducing landfill burden.

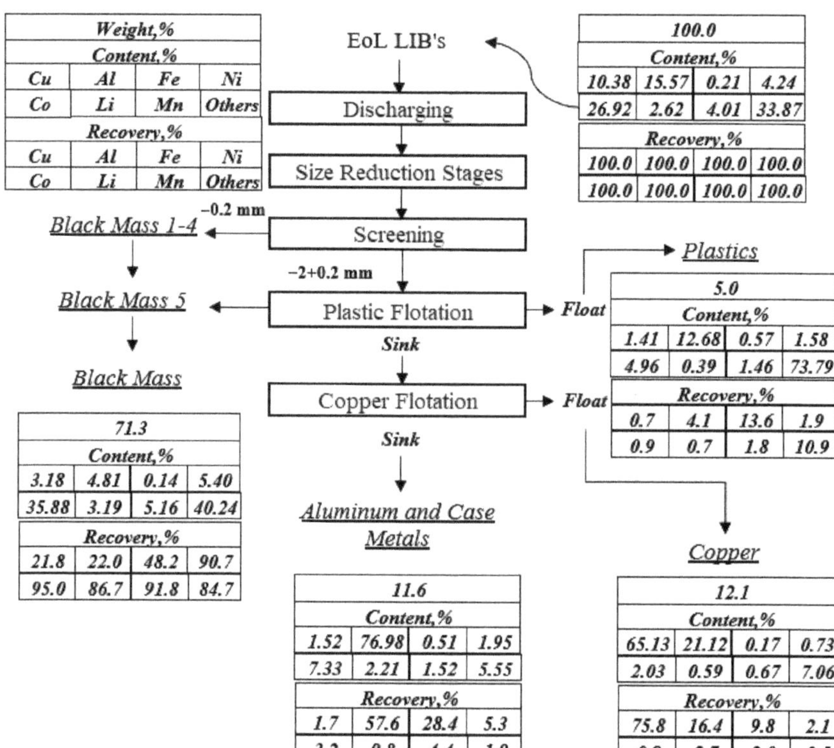

Figure 8. Proposed flowchart for recycling of EoL LiBs by the flotation method.

The non-floated fraction was further processed in a Cu flotation circuit using 3000 g/t 3418A as the collector. This step produced a Cu concentrate representing 12.1% of the feed mass, assaying 65.13% Cu with 96.4% recovery, which constitutes a high-value feedstock for downstream hydrometallurgical or pyrometallurgical refining and supports efficient circular use of critical metals. The remaining tailings, constituting 11.6% of the initial feed and containing 76.98% Al, represent a secondary aluminum source that can be further upgraded by electrostatic or eddy current separation, potentially reducing reliance on primary aluminum production and its associated energy consumption.

The observed decrease in overall Cu recovery is primarily attributed to the removal of fine BM during the crushing and classification stages, which inevitably contains a fraction of fine Cu and Al current collector particles. To minimize such losses and improve the overall process efficiency, future work could focus on implementing additional pretreatment steps aimed at selectively scraping off the active electrode materials before size reduction. Potential approaches include mild thermal treatment to weaken the adhesion of the electrode coatings or controlled mechanical abrasion techniques to remove the BM without causing excessive fragmentation of the metallic foils.

To enhance Cu content and Al recovery, it is also recommended to reduce the floatability of Al particles through Sodium Hydroxide (NaOH) pre-treatment, followed by feeding them into a cleaning flotation circuit. Alternatively, reshaping the particles into more spherical forms via appropriate grinding techniques may facilitate the separation of Al and Cu using electrostatic or eddy current separation methods. Moreover, the Al-rich sink product obtained after Cu flotation, which remains unfloated due to its low hydrophobicity, can be further processed through magnetic separation to remove associated ferromagnetic metals. This step not only increases the purity of the Al fraction but also enables the production of a marketable magnetic product enriched in Fe and Ni.

Compared with conventional separation methods, the proposed two-step flotation process offers several advantages. Mechanical screening and density-based classification are generally less effective at liberating and separating plastics from metallic foils, often resulting in residual plastic content exceeding 15 wt.% in the metallic fraction [40]. In contrast, our process achieved more than 95% plastic removal, producing a clean Cu–Al feed for subsequent processing. It should be noted that the collector dosage applied in this study (3×1000 g/t) is higher than typical values used in conventional mineral flotation. This is attributed to the presence of binder residues, carbon coatings, and passive surface layers on the Cu foils, which require more reagents to achieve effective surface hydrophobization.

The approach used in this study can be contrasted with that of Saneie et al. [32], who investigated froth flotation for the recovery of Cu and Al from 18650-type cylindrical LiBs using a 1 mm cut size. Under such conditions, a substantial fraction of current collector foils is lost with the BM, reducing overall metal recovery potential. In the present work, a finer cut size of 0.2 mm was selected specifically to minimize these losses. Moreover, a multi-stage crushing and classification stage was applied to remove approximately 95% of the residual BM adhering to the foils before flotation. This combination of finer classification and pre-cleaning not only improves the selectivity of the subsequent flotation stages but also makes the process more representative of industrial practice for mixed NMC and LCO batteries from electric vehicles, where maximizing metal recovery and minimizing contamination of the plastic fraction are crucial objectives.

On a per-ton basis, the reagent usage of the proposed flowsheet is as follows: the Cu stage collector (3418A) corresponds to $\approx$3.0 kg/t feed (3×1000 g/t), while the frother (MIBC) consumption at the selected operating point is $\approx$0.5 kg/t. Based on typical procurement prices for laboratory and industrial use (3418A: \$10–30/kg [41]; MIBC: \$3–10/kg [42]), the total reagent cost is estimated at approximately \$45–105 per ton of feed, with MIBC contributing around \$1.5–5/t. According to the corrected mass balance, the Cu concentrate accounts for 12.1 wt.% of the feed at 65.13 wt.% Cu, corresponding to ~0.121 t concentrate/t feed carrying ~0.079 t Cu. The Al-rich sink assays ~77 wt.% Al. By removing plastics and fines before chemical processing, the pre-concentration step is expected to reduce acid consumption and furnace duty, thereby lowering operating costs and greenhouse-gas emissions relative to direct treatment of mixed feed [43].

From a scale-up perspective, the reagent consumptions reported here are within a range that is technically feasible for pilot- and industrial-scale operations, and the flowsheet can be directly adapted to continuous flotation circuits with minimal modification. Future work should include a pilot-scale verification to validate reagent dosages and recovery efficiencies under steady-state conditions.

By implementing flotation as a pre-concentration step, the amount of inert material entering downstream processes can be significantly reduced, leading to improved leaching kinetics, lower reagent consumption, and reduced energy demand during thermal steps. These results indicate that flotation-based pre-treatment is both technically and economically advantageous compared to conventional approaches.

4. Conclusions

In this study, more than 95% of the plastic fraction was removed in the first flotation stage using MIBC as a frother, producing a low-metal discard stream. In the second stage, Cu foils were selectively floated with 3418A, yielding a Cu concentrate assaying 65.13% Cu with 96.4% recovery. The remaining tailings were enriched in Al (~77%), representing a valuable secondary feed material for further upgrading using electrostatic, eddy current, or magnetic separation methods. These results confirm that plastics, Cu, and Al can be efficiently liberated and separated through appropriate pre-treatment and flotation strategies.

Beyond demonstrating the technical feasibility of this approach, the study also highlights several process optimization opportunities. The partial loss of fine Al particles to the Cu concentrate suggests that pre-conditioning steps such as NaOH washing or particle reshaping could further enhance Al recovery and concentrate purity. Integration of scavenger flotation or post-flotation physical separation could reduce entrained plastics and residual active material, further improving overall process efficiency.

While these results are promising, some limitations remain. Additional replicate flotation tests and statistical analysis are needed to confirm the robustness of the reported recoveries. Future work will explore alkaline pre-treatment (e.g., NaOH washing) or mechanical/thermal surface-cleaning methods to further reduce electrode coating contamination, as well as the integration of this approach with downstream hydrometallurgical processes for complete metal recovery. Finally, pilot-scale experiments should be performed to evaluate process scalability and economic feasibility under realistic industrial conditions, ensuring that the proposed approach can be effectively implemented in LiBs recycling plants and contribute to a more sustainable circular economy for critical metals.

Author Contributions: Conceptualization, F.B.; methodology, F.M., N.İ.D. and F.B.; validation, F.M., F.B., Z.Š. and R.L.; formal analysis, B.T.K. and R.L.; investigation, F.M., N.İ.D. and F.B.; resources, F.B.; data curation, F.M. and F.B.; writing—original draft preparation, F.M., F.B., Z.Š., R.L. and B.T.K.; writing—review and editing, F.B., Z.Š., R.L. and B.T.K.; visualization, F.M., F.B. and R.L.; supervision, F.B.; project administration, F.M. and F.B.; funding acquisition, F.M. and F.B. All authors have read and agreed to the published version of the manuscript.

Funding: The present study is based on the results of the graduate thesis conducted at the Istanbul Technical University and supported by Istanbul Technical University, BAP (Scientific Research Project) Department, with the project codes MGA-2022-43640 and MYL-2023-44587.

Data Availability Statement: The original contributions presented in this study are included in the article. Further inquiries can be directed to the corresponding author.

Acknowledgments: The authors sincerely thank the Mineral Processing Department for providing laboratory equipment and Exitcom Company for the supply of samples.

Conflicts of Interest: The authors declare that the research was conducted in the absence of any commercial or financial relationships that could be construed as a potential conflict of interest.

Abbreviations

The following abbreviations are used in this manuscript:

EoL	End of Life
LiBs	Lithium-Ion Batteries
Cu	Copper
Al	Aluminum
Li	Lithium
Co	Cobalt
Ni	Nickel
Fe	Iron
Mn	Manganese
C	Carbon
P	Phosphorus
H	Hydrogen
S	Sulfur
NaOH	Sodium Hydroxide
EVs	Electric Vehicles
BESS	Battery Energy Storage Systems
PVDF	Polyvinylidene Fluoride
LCO	lithium cobalt oxide
NMC	Nickel Cobalt Manganese Oxide
NCA	Nickel Cobalt Aluminum Oxide
LMO	Lithium Manganese Oxide
LFP	Lithium Iron Phosphate
LTO	Lithium Titanate Oxide
BM	Black Mass
PCBs	Printed Circuit Boards
KAX	Potassium Amyl Xanthate
MIBC	Methyl Isobutyl Carbinol
XRF	X-Ray Fluorescence Spectroscopy
ICP	Inductively Coupled Plasma
SEM	Scanning Electron Microscopy
EDS	Energy Dispersive Spectroscopy
SE	Secondary Electron
BSE	Backscattered Electron
LOI	Loss on Ignition

References

1. Kaya, M. State-of-the-art lithium-ion battery recycling technologies. *Circ. Econ.* **2022**, *1*, 100015. [CrossRef]
2. Zhang, Y.; Meng, Q.; Dong, P.; Duan, J.; Lin, Y. Use of grape seed as reductant for leaching of cobalt from spent lithium-ion batteries. *J. Ind. Eng. Chem.* **2018**, *66*, 86–93. [CrossRef]
3. Meshram, P.; Mishra, A.; Sahu, R. Environmental impact of spent lithium ion batteries and green recycling perspectives by organic acids—A review. *Chemosphere* **2020**, *242*, 125291. [CrossRef] [PubMed]

4. Mylarappa, M.; Lakshmi, V.V.; Mahesh, K.V.; Nagaswarupa, H.P.; Prashantha, S.C.; Kumara, K.S.; Raghavendra, N. Resource recovery and material characterization of metals from waste li-ion batteries by an eco-friendly leaching agent. *Mater. Today* **2017**, *4*, 12215–12222. [CrossRef]
5. Li, D.; Gao, Y.; Chen, Y.; Peng, E.; Yu, D.; Guo, X. Green and efficient combined pretreatment for enhanced cathode hydrophilicity and flotation separation in spent lithium-ion batteries. *J. Sep. Sci.* **2025**, *367*, 132862. [CrossRef]
6. Metal Home Page. Available online: https://news.metal.com/newscontent/103286919/Key-Takeaway-from-CLNB-2025-10th-New-Energy-Industry-Chain-Expo%3A%C2%A0Current-Status-and-Outlook-of-Chinas-Lithium-Ion-Battery-Industry%3A-How-Can-Enterprises-Survive-Cut-Throat-Competition%C2%A0 (accessed on 29 July 2025).
7. Xu, J.; Thomas, H.R.; Francis, R.B.; Lumb, K.R.; Wanga, J.; Liang, B. A review of processes and technologies for the recycling of lithium-ion secondary batteries. *J. Power Sources* **2008**, *177*, 512–527. [CrossRef]
8. Zhang, T.; He, Y.; Wang, F.; Li, H.; Duan, C.; Wu, C. Surface analysis of cobalt-enriched crushed products of spent lithium-ion batteries by X-ray photoelectron spectroscopy. *Separ. Purif. Technol.* **2014**, *138*, 21–27. [CrossRef]
9. Mennik, F.; Burat, F. The effect of attrition scrubbing as a pre-treatment on the recovery of copper from end-of-life lithium-ion batteries in froth flotation. In Proceedings of the 28th International Mining Congress and Exhibition of Türkiye, Antalya, Turkey, 28 November–1 December 2023.
10. Pillot, C. The Rechargeable Battery Market and Main Trends 2014–2025. Adv. Automot. Batter. Conf. 2016. Available online: http://cii-resource.com/cet/FBC-TUT8/Presentation/Pillot_Christophe.pdf (accessed on 27 July 2025).
11. Wang, C.; Tao, R.; Xing, P.; Ni, J.; Cun, Z.; Liu, S.; Wang, X.; Dai, Y.; Long, Z.; Sun, Z.; et al. State-of-the-art review on liberation of cathode material in pretreatment recycling of spent lithium-ion battery. *Resour. Conserv. Recycl.* **2025**, *218*, 108257. [CrossRef]
12. Fu, Y.; He, Y.; Li, J.; Qu, L.; Yang, Y.; Guo, X.; Xie, W. Improved hydrometallurgical extraction of valuable metals from spent lithium-ion batteries via a closed-loop process. *J. Alloys Compd.* **2020**, *847*, 156489. [CrossRef]
13. Zhang, T.; He, Y.; Wang, F.; Ge, L.; Zhu, X.; Li, H. Chemical and process mineralogical characterizations of spent lithium-ion batteries: An approach by multi-analytical techniques. *J. Waste Manag.* **2014**, *34*, 1051–1058. [CrossRef]
14. Gratz, E.; Sa, Q.; Apelian, D.; Wang, Y. A closed loop process for recycling spent lithium ion batteries. *J. Power Sources* **2014**, *262*, 255–262. [CrossRef]
15. Meshram, P.; Pandey, B.D.; Mankhand, T.R. Recovery of valuable metals from cathodic active material of spent lithium ion batteries: Leaching and kinetic aspects. *J. Waste Manag.* **2015**, *45*, 306–313. [CrossRef]
16. Tong, Z.; Wang, M.; Bai, Z.; Li, H.; Wang, N. Advances in lithium-ion battery recycling: Strategies, pathways, and technologies. *ChemPhysMater* **2025**, *4*, 30–47. [CrossRef]
17. Asadi Dalini, E.; Karimi, G.; Zandevakili, S.; Goodarzi, M. A Review on Environmental, Economic and Hydrometallurgical Processes of Recycling Spent Lithium-ion Batteries. *Miner. Process. Extr. Metall. Rev.* **2021**, *42*, 451–472. [CrossRef]
18. Ma, X.; Meng, Z.; Bellonia, M.V.; Spangenberger, J.; Harper, G.; Gratz, E.; Olivetti, E.; Arsenault, R.; Wang, Y. The evolution of lithium-ion battery recycling. *Nat. Rev. Clean Technol.* **2025**, *1*, 75–94. [CrossRef]
19. Kim, S.; Bang, J.; Yoo, J.; Shin, Y.; Bae, J.; Jeong, J.; Kim, K.; Dong, P.; Kwon, K. A comprehensive review on the pretreatment process in lithium-ion battery recycling. *J. Clean. Prod.* **2021**, *294*, 126329. [CrossRef]
20. Liao, H.; Zhang, S.; Liu, B.; He, X.; Deng, J.; Ding, Y. Valuable metals recovery from spent ternary lithium-ion battery: A review. *Int. J. Miner. Metall. Mater.* **2024**, *31*, 2556–2581. [CrossRef]
21. Yao, L.; Yao, H.; Xi, G.; Feng, Y. Recycling and synthesis of LiNi1/3Co1/3Mn1/3O2 from waste lithium ion batteries using d,l-malic acid. *RSC Adv.* **2016**, *6*, 17947–17954. [CrossRef]
22. Zhang, T.; He, Y.; Ge, L.; Fu, R.; Zhang, X.; Huang, Y. Characteristics of wet and dry crushing methods in the recycling process of spent lithium-ion batteries. *J. Power Sources* **2013**, *240*, 766–771. [CrossRef]
23. Bertuol, D.A.; Toniasso, C.; Jiménez, B.M.; Meili, L.; Dotto, G.L.; Tanabe, E.H.; Aguiar, M.L. Application of spouted bed elutriation in the recycling of lithium ion batteries. *J. Power Sources* **2015**, *275*, 627–632. [CrossRef]
24. Shi, J.; Peng, C.; Chen, M.; Li, Y.; Eric, H.; Klemettinen, L.; Lundström, M.; Taskinen, P.; Jokilaakso, A. Sulfation Roasting Mechanism for Spent Lithium-Ion Battery Metal Oxides Under SO2-O2-Ar Atmosphere. *J. Miner.* **2019**, *71*, 4473–4482. [CrossRef]
25. Segura-Bailón, B.; Lapidus, G.T.; Ramos-Sánchez, G. A comparative study of discharging and leaching of spent lithium-ion battery recycling. *Miner. Eng.* **2024**, *218*, 109012. [CrossRef]
26. Shen, J.; Zhou, M.; Liu, W.; Shi, Y.; Tang, W.; Deng, Y.; Liu, R.; Zuo, Y.; Zhang, J. Advanced direct recycling technology enables a second life of spent lithium-ion battery. *Energy Storage Mater.* **2025**, *74*, 103964. [CrossRef]
27. Chen, X.; Guo, C.; Ma, H.; Li, J.; Zhou, T.; Cao, L.; Kang, D. Organic reductants based leaching: A sustainable process for the recovery of valuable metals from spent lithium ion batteries. *J. Waste Manag.* **2018**, *75*, 459–468. [CrossRef]
28. Aarab, I.; Amari, K.E.; He, D.; Fu, Y.; Boujounoui, K.; Etahiri, A. Eco-Friendly apatite flotation: Unlocking the potential of spent coffee grounds as a bio-based collector. *Miner. Eng.* **2025**, *233*, 109652. [CrossRef]
29. Nazari, S.; Vakylabad, A.B.; Asgari, K.; Li, J.; Khoshdast, H.; He, Y.; Hassanzadeh, A. Bubbles to batteries: A review of froth flotation for sustainably recycling spent lithium-ion batteries. *J. Energy Storage.* **2024**, *84*, 110702. [CrossRef]

30. Arellano-Sanchez, D.; Rinne, M.; Gomez-Moreno, L.A.; Wilson, B.P.; Serna-Guerrero, R.; Lundström, M. Life cycle assessment of enhanced materials recovery from LTO-rich anode waste recycling via froth flotation. *Miner. Eng.* **2025**, *227*, 109271. [CrossRef]
31. Liu, J.; Wang, H.; Hu, T.; Bai, X.; Wang, S.; Xie, W.; Hao, J.; He, Y. Recovery of $LiCoO_2$ and graphite from spent lithium-ion batteries by cryogenic grinding and froth flotation. *Miner. Eng.* **2020**, *148*, 106223. [CrossRef]
32. Saneie, R.; Abdollahi, H.; Ghassa, S.; Azizi, D.; Chehreh Chelgani, S. Recovery of Copper and Aluminum from Spent Lithium-Ion Batteries by Froth Flotation: A Sustainable Approach. *J. Sustain. Metall.* **2022**, *8*, 386–397. [CrossRef]
33. Yenial, U.; Burat, F.; Yüce, A.E.; Güney, A.; Kangal, M.O. Separation of PET and PVC by flotation technique without using alkaline treatment. *Miner. Process. Extr. Metall. Rev.* **2013**, *34*, 412–421. [CrossRef]
34. Güney, A.; Özdilek, C.; Kangal, M.O.; Burat, F. Flotation characterization of PET and PVC in the presence of different plasticizers. *Sep. Purif. Technol.* **2015**, *151*, 47–56. [CrossRef]
35. Dinç, N.İ.; Tosun, A.U.; Baştürkcü, E.; Özer, M.; Burat, F. Recovery of valuable metals from WPCB fines by centrifugal gravity separation and froth flotation. *J. Mater. Cycles Waste Manag.* **2022**, *24*, 224–236. [CrossRef]
36. Kikongi, P.; Salvas, J.; Gosselin, R. Curve-fitting regression: Improving light element quantification with XRF. *X-Ray Spectrom.* **2017**, *46*, 347–355. [CrossRef]
37. Solvay. Aerophine® 3418A Collector—Technical Data Sheet. 2021. Available online: https://www.solvay.com (accessed on 5 May 2021).
38. 911Metallurgist. Aerophine 3418A Collector: Performance & Chemistry. Available online: https://www.911metallurgist.com/blog/aerophine-3418a/ (accessed on 25 May 2025).
39. Fadel, M.A.; Kamel, N.A.; Darwish, M.M.; El-Messieh, S.L.A.; Abd-EL-Nour, K.N.; Khalil, W.A. Preparation and characterization of polyethylene terephthalate–chamomile oil blends with enhanced hydrophilicity and anticoagulant properties. *Prog. Biomater.* **2020**, *9*, 97–106. [CrossRef] [PubMed]
40. Mennik, F.; Dinç, N.İ.; Burat, F. Selective recovery of metals from spent mobile phone lithium-ion batteries through froth flotation followed by magnetic separation procedure. *Results Eng.* **2023**, *17*, 100868. [CrossRef]
41. Clariant Mining Solutions. Aerophine® 3418A Collector-Product Data Sheet. Clariant. 2023. Available online: https://www.clariant.com/mining (accessed on 10 September 2025).
42. ICIS Chemical Business. Methyl Isobutyl Carbinol (MIBC) Global Market Price Trends. ICIS. 2022. Available online: https://www.icis.com/chemicals/mibc (accessed on 10 September 2025).
43. Burat, F.; Güney, A.; Kangal, M.O. Selective separation of virgin and post-consumer polymers (PET and PVC) by flotation method. *J. Waste Manag.* **2009**, *29*, 1807–1813. [CrossRef]

minerals

MDPI

Article

Study on the Inhibition and Activation of Pyrite Under Low Alkalinity Conditions Created by Hydrogen Peroxide and Lime

Yuankun Yang [1,2,*], Milena Kostović [1,*], Rongdong Deng [3] and Yinying Liao [2,3]

1 Faculty of Mining and Geology, University of Belgrade, 11000 Belgrade, Serbia
2 Zijin Mining Group Co., Ltd., Longyan 364200, China; liao_yinying@zijinmining.com
3 Zijin School of Geology and Mining, Fuzhou University, Fuzhou 350108, China; dengrongdong111@163.com
* Correspondence: yang.yuankun@dok.rgf.bg.ac.rs (Y.Y.); milena.kostovic@rgf.bg.ac.rs (M.K.)

Abstract

High alkalinity facilitates copper–sulfur flotation separation but also leads to issues such as high reagent consumption, pipeline scaling, and gold loss in tailings. The ore from a copper mine in Serbia contains 2.86% copper, 1.64 g/t gold, and 20.39% sulfur, with copper occurring mainly in covellite and enargite. To achieve efficient separation and recovery of copper–sulfur, a systematic study was conducted using micro-flotation, Scanning Electron Microscopy (SEM), X-ray photoelectron spectroscopy (XPS) and contact angle analysis to investigate the inhibition and activation patterns of pyrite under low and high alkalinity conditions. The results indicate that the combined use of hydrogen peroxide and lime as inhibitor enables efficient separation of pyrite and covellite under low-alkalinity conditions. This effect is attributed to its ability to enhance oxidation of the pyrite surface, which generates more hydrophilic substances. Under low-alkalinity conditions (slurry pH = 10) regulated with hydrogen peroxide and lime in a covellite flotation cycle, and under acidic conditions (slurry pH = 6) in the pyrite flotation cycle, satisfactory results are obtained in both flotation cycles in comparison with industrial data. The copper flotation index was similar, but pyrite and gold recovery increased by 2.3% and ~4%, respectively, over those using lime alone. This process reduced the activator dosage required for pyrite activation substantially, while improving gold recovery. Results demonstrate an efficient method for copper–sulfur separation and recovery, providing theoretical guidance or industrial production processes.

Keywords: copper-sulfur separation; low-alkalinity; hydrogen peroxide; flotation; covellite; pyrite

1. Introduction

Pyrite (FeS_2) is a common iron sulfide mineral and an important raw material for the preparation of sulfuric acid (H_2SO_4) and sulfur [1–4]. Covellite (CuS), an important raw material for copper smelting, is a widespread copper sulfide mineral accounting for approximately 70% of the world's total copper reserves [5,6]. The intergrowth or association of covellite and pyrite is extremely common in nature [7]. The natural floatability of covellite is similar to but slightly higher than that of pyrite [8–10]. Under low-alkalinity conditions, efficient separation of the two minerals is difficult to achieve. Currently, the most commonly used method for effective separation is the inhibition of pyrite flotation in a high-alkalinity environment, which increases the difference in floatability between the two [11,12].

The efficient separation and recovery of pyrite and covellite is essential for resource utilization. Commonly used pyrite inhibitors include lime, cyanide, sulfur oxides, and oxidizing agents [13,14]. Lime inhibitors inhibit the flotation recovery of pyrite by adjusting the slurry pH. A higher pH increases the OH^- concentration and the hydrophilic hydroxide content on the surface of pyrite, thereby inhibiting the flotation recovery of pyrite. Although lime possesses many advantages in pyrite inhibition, the excessive lime addition often leads to scaling, solidification, and pipeline blockage. The dosage of lime is difficult to control, and an excessively high slurry alkalinity leads to high collector consumption, which is not favorable for comprehensive resource recycling. Many researchers have thus focused on inhibitors of pyrite in low-alkalinity media [15,16]. For pyrite containing the precious metal gold, floatation conditions also significantly influence gold recovery.

Oxidizing agents are attractive alternatives due to their wide availability, low cost, and good selectivity. They also exert a smaller impact on associated valuable metals and are suitable for use over a wider slurry pH range [17]. Therefore, they are widely applied in copper–sulfur separation. Hydrogen peroxide (H_2O_2), a clean and efficient oxidant with low human toxicity, has recently become more widely used in mineral processing. Khoso et al. [18] achieve the selective separation of chalcopyrite and pyrite using H_2O_2 at a pH of 9.0, with the recovery of chalcopyrite exceeding 84% and that of pyrite being lower than 24%. Infrared and X-ray photoelectron spectroscopy (XPS) analyses revealed that H_2O_2 treatment substantially altered the surface properties of pyrite, rendering it highly hydrophilic and inhibiting collector adsorption, thereby reducing its floatability. Wang et al. [19] have shown that the surface oxidation of pyrite is increased and dixanthogen formation is decreased under high-alkalinity slurry flotation. This leads to the reduced adsorption of xanthate, which in turn inhibits the flotation of pyrite. Ozun et al. [20] have reported that pyrite is a chemically active mineral whose surface oxidizes when exposed to the atmosphere and aqueous environments during mining and flotation. The oxidation process results in the creation of a hydrophilic surface, which affects collector adsorption and flotation effects. Luo et al. [21] have regulated the dissolved oxygen concentration in slurry by adjusting the relative proportion of O_2/N_2 during aeration. The generation of dixanthogen and avoidance of excessive iron oxide/hydroxide formation on the pyrite surface contribute to the regulation of pyrite recovery by flotation. Copper flotation is associated with various issues, such as a low ore grade, high pyrite content, and complex mineral dissemination relationships [22,23]. For pyrite, a carrier mineral for common gold monomers, the effects of cyanide and hydroxides must be minimized. In other words, efficient recovery of elemental gold can only proceed in low-alkalinity environments [24]. Yuan [25], Yang [26], and Khosol et al. [18] have systematically evaluated low-alkalinity flotation systems. Besides inhibiting the flotation performance of pyrite by adjusting the environment of the flotation slurry to low alkalinity, the researchers also introduced oxidizing agents to further enhance the inhibitory effect. In addition to oxidation treatment in a low-alkalinity solution environment, the oxidizing agent dosage was reduced to 1/3 of the original dosage. The efficient formation of a passivation layer on the pyrite surface was achieved, which significantly inhibited floatability and effectively reduced the sulfur grade by 10%–20% in the copper concentrate.

In this paper, the depression and activation mechanisms of pyrite under high-alkali and low-alkali conditions were systematically investigated. The research findings were successfully applied at a mine in Serbia, achieving effective copper and sulfur copper–sulfur separation and recovery. The low-alkali process significantly improved the production indicators of the beneficiation plant while reducing reagent consumption, providing valuable insights for similar mining operations.

2. Materials and Methods

2.1. Ore Sample

The main chemical composition, particle size distribution and liberation relationship analysis of ore sample are presented in Tables 1–3.

Table 1. Main Chemical Composition of Ore from the Upper Zone of a Copper-Gold Mine in Sebia (Unit: %, *: g/t).

Elements	Cu	Au *	S	As	Fe	Pb	SO_3	Zn	SiO_2
Contents	2.8	1.64	20.39	0.035	18.87	0.01	3.74	0.02	3.74

As shown in Table 1, the main valuable elements in the raw ore were copper, gold, and sulfur. Copper and gold grades were 2.8% and 2.16 g/t, respectively, whereas sulfur possessed a lower value.

Tables 2 and 3 show the mineralogical results of the ore sample. As shown in Table 2, 85.88% of the copper sulfides was completely liberated at a particle size of 70% passing 75 μm. Rich intergrowths with 50% ≤ X < 100% accounted for 10.08%, whereas lean intergrowths (X < 50%) accounted for only 4.04%.

Results shown in Table 3 indicate that at a particle size of 70% passing 75 μm, 83.77% of the pyrite was completely liberated, with the vast majority being in the floatable particle size range. Rich intergrowths accounted for 6.12% of the pyrite. It is noted that the pyrite in <10 μm size fraction that is difficult to recover by flotation accounted for only 1.17%. Lean intergrowths accounted for 6.74% of the pyrite. Floatable pyrite accounted for 92.99% of the total pyrite (>10 μm and X > 50%).

At a particle size of 70% passing 75 μm, the concentrate obtained by gravity concentration was examined manually using an optical stereomicroscope (Nikon Corporation, Tokyo, Japan) and subjected to MLA testing. Six liberated native gold particles with a particle size of up to 17 μm were observed, with no encapsulated gold observed. Figure 1 shows the concentrated liberated native gold particles.

Figure 1. Liberated Gold in Raw Ore (Optical Microscope Image).

Table 2. Particle Size Distribution and Liberation Relationship Analysis of Copper Sulfides in Ore.

Copper–Sulfur Mineral Intergrowth Type	Yield (%)	Intergrowth Association		Copper Sulfide Particle Size Distribution (%)					
		With Py	With Others	−10 μm	−20 + 10 μm	−38 + 20 μm	−74 + 38 μm	−150 + 74 μm	+150 μm
X = 100%	85.88	--	--	0.62	14.7	26.15	30.03	14.12	0.26
80% ≤ X < 100%	6.22	4.34	1.88	--	0.12	0.77	2.26	2.81	0.26
50% ≤ X < 80%	3.86	3.29	0.57	--	0.24	1.47	1.74	0.41	--
X < 50%	4.04	3.07	0.97	0.6	1.04	1.69	0.71	--	--
Sum	100.00	10.7	3.42	1.22	16.1	30.08	34.74	17.34	0.52

Note: X represents the proportion of the target mineral in the liberation particle; "Py" is the abbreviation for pyrite, used consistently throughout the text.

Table 3. Particle Size Distribution and Liberation Relationship Analysis of Pyrite.

Type of Pyrite Intergrowth	Yield (%)	Intergrowth Association		Pyrite Particle Size Distribution (%)					
		With Cu	With Others	−10 μm	−20 + 10 μm	−38 + 20 μm	−74 + 38 μm	−150 + 74 μm	+150 μm
X = 100%	83.77	--	--	0.23	18.47	30.35	27.41	7.22	0.09
80% ≤ X < 100%	6.12	2.08	4.04	0.01	0.24	1.43	3.06	1.38	--
50% ≤ X < 80%	3.37	0.58	2.79	0.03	0.49	0.96	1.73	0.16	--
X < 50%	6.74	0.63	6.11	0.90	2.22	2.51	1.10	0.01	--
Sum	100.00	3.29	12.94	1.17	21.42	35.25	33.30	8.77	0.09

The results of a chemical phase analysis are summarized in Table 4. Covellite is the main copper sulfide mineral contained in the ore sample, and the major gangue minerals include pyrite and quartz, with pyrite being the primary gold carrier. Therefore, it is necessary to perform the efficient separation and recovery of copper and sulfur.

Table 4. Chemical Phase Analysis of Serbia Gold–Copper Ore.

Mineral Species	Mineral Content/%
Covellite (CuS)	11.49
Tennantite (Cu_3AsS_4)	0.89
Pyrite (FeS_2)	39.31
Quartz (SiO_2)	27.74

2.2. Flotation Process of the Mine

At present, the preferential flotation process is adopted at processing plant (Figure 2). Lime is used as the pyrite inhibitor, Z200 is used as the collector, and the slurry pH is maintained at approximately 12. Current flotation indicators on the site are shown in Table 5, the obtained copper concentrate has a copper grade of 21.88%, copper recovery of 94.38%, and gold recovery of 32.23%. The sulfur concentrate has a sulfur grade of 49.40%, recovery of 56.70%, and gold recovery of 53.87%. Major production issues are as follows: (1) high dosage of lime (approximately 5 kg/t), resulting in a sticky flotation foam and a high tendency for slurry pipeline blockage, and (2) low sulfur flotation recovery rate and low gold recovery during sulfur flotation. A low-alkalinity process, in which lime is partially replaced by an oxidizing agent, is an effective method to reduce the amount of lime used while increasing the recovery of gold during sulfur separation.

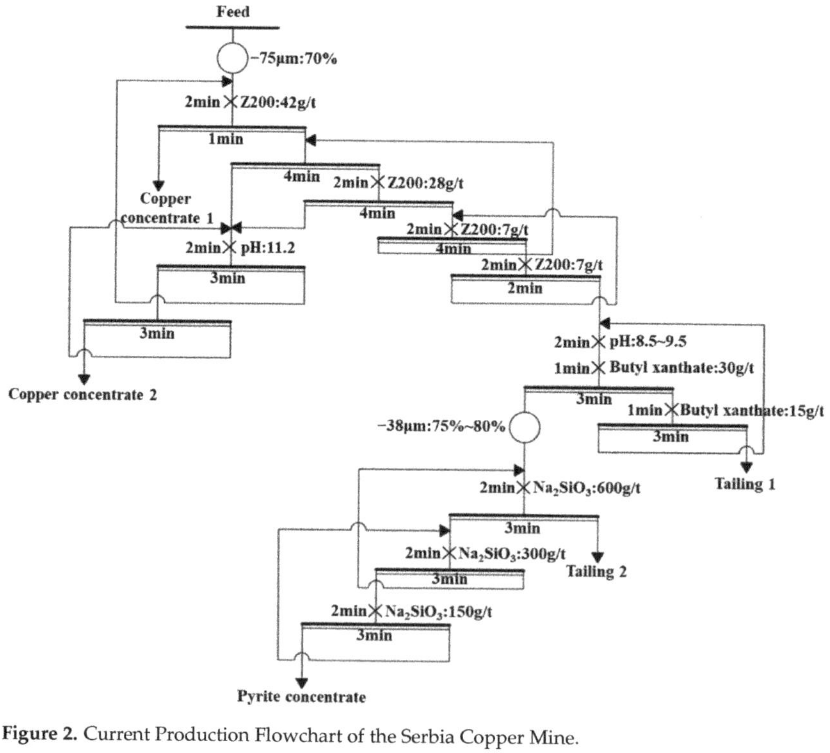

Figure 2. Current Production Flowchart of the Serbia Copper Mine.

Table 5. Current Flotation indicators of the Serbia Copper Mine.

Product	Yield (%)	Grade (%)			Recovery (%)		
		Cu	Au ($g{\cdot}t^{-1}$)	S	Cu	Au	S
Copper Concentrate 1	7.57	30.35	4.07	34.70	80.42	18.81	12.89
Copper Concentrate 2	4.76	8.39	4.62	36.80	13.97	13.42	8.58
Pyrite Concentrate	23.41	0.52	3.77	49.40	4.26	53.87	56.70
Tailing 2	14.53	0.096	0.78	7.61	0.49	6.92	5.42
Tailing 1	49.73	0.050	0.23	6.73	0.87	6.98	16.41
Feed	100.00	2.86	1.64	20.39	100.00	100.00	100.00
Copper Concentrate 1 + 2	12.33	21.88	4.28	35.51	94.38	32.23	21.47
Tailing 1 + 2	64.26	0.06	0.35	6.93	1.36	13.90	21.83

The tests were performed with pure pyrite mineral and raw ore sample from upper zone of the deposit from a cooper mine in Serbia.

2.3. Pure Mineral

Figure 3 shows the X-ray diffraction (XRD) spectrum of the pure pyrite mineral sample used as the raw material for single-mineral flotation testing. As shown in Figure 3, the spectrum is free of any obvious impurity peaks. Together with chemical analyses, it revealed that the purity of pyrite exceeded 98%; the spectrum indicates that the pyrite sample fulfills the requirement for single mineral flotation tests.

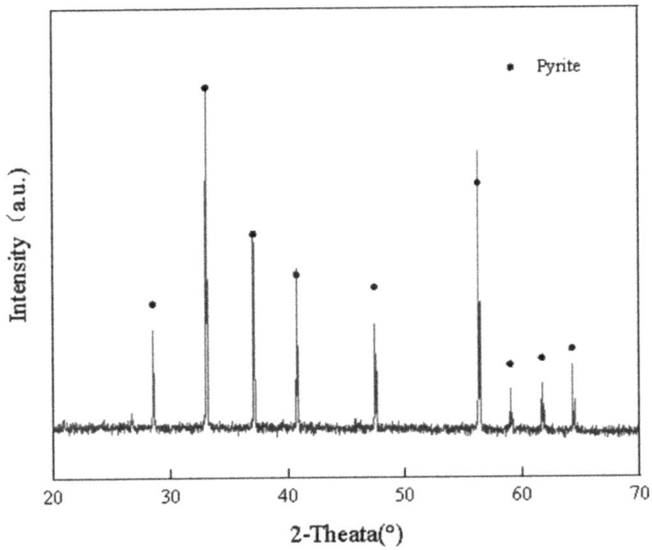

Figure 3. XRD Pattern of Pure Pyrite Mineral.

2.4. Reagents

Table 6 shows the reagents used in this study, including regulators, collectors, inhibitors, and frothers. To note, H_2SO_4, when used, was diluted to a 1% solution with distilled water.

Table 6. Main Chemical Reagents Used in the Experiment.

Chemical Formula	Content (wt.%)	Supplier	Function
CaO	98	Macklin Biochemical (Shanghai), Shanghai, China	Depressant
H_2SO_4	98.08	Sinopharm Chemical Reagent Co., Ltd., Shanghai, China	pH regulator
NaBX	—	Zhuzhou Flotation Reagents Co., Ltd., Zhuzhou, China	collector
MIBC	99	Shanghai EON Chemical Technology Co., Shanghai, China	frother
Z200($C_3H_7CH(CH_3)C(S)NHC_2H_5$)	—	Zhuzhou Flotation Reagents Co., Ltd., Zhuzhou, China	collector
H_2O_2	30	Macklin Biochemical (Shanghai), Shanghai, China	depressant

2.5. Methods and Processes

Pure mineral flotation testing was performed using a hanging cell flotation machine(Shandong Xinhai Mining Technology & Equipment Co., Ltd., Yantai, China). During each test, 2.0 g of the mineral sample was weighed and rapidly put into the 40 mL hanging cell flotation machine using a solution at the corresponding pH. The slurry was stirred for 3 min, and the slurry pH was measured. Subsequently, the regulator, collector, and frother were added sequentially, and the slurry was mixed and reacted for 2, 2, and 1 min. Aerated flotation was performed for 3 min, and the concentrate and tailings were dried in an oven and weighed. The recovery rate (%) was calculated using the following equation:

$$\varepsilon = \frac{m_1}{m_1 + m_2} \times 100 \tag{1}$$

ε—recovery rate (%),
m_1—the mass of the froth product (concentrate weight, g),
m_2—the mass of the in-cell product (tailing weight, g).

Raw ore flotation testing was performed using laboratory single-cell flotation machines (Shandong Xinhai Mining Technology & Equipment Co., Ltd., Yantai, China) with capacities of 1.5, 1, and 0.5 L. For each test, 500 g of the ore sample was weighed, and copper concentrate flotation and separation testing was performed based on the optimum reagent system and test procedure derived from single-mineral pyrite flotation testing.

2.6. Testing and Analysis

2.6.1. Scanning Electron Microscopy (SEM)

The surface morphology of pure natural pyrite samples was analyzed using SEM (FEI Quanta 200, FEI Company, Hillsboro, OR, USA). Pyrite samples (2 g, particle size range: −74 + 45 μm) were treated using the reagent addition sequence and reaction times used in flotation testing, subjected to solid–liquid separation, and vacuum-dried at a low temperature before use as a test material.

2.6.2. Contact Angle Measurement

The wettability of mineral surfaces was characterized by measuring the contact angle using a drop shape analyzer (DSA255; KRŰSS, Hamburg, Germany). For sample preparation, an appropriately sized pyrite sample block was obtained, and one surface was ground sequentially until smooth with coarse to fine sandpaper. Subsequently, surface impurities were washed with dilute sulfuric acid, and the surface of the pyrite sample was rinsed with deionized water. The sample was then treated using the reagent addition sequence and reaction times used in flotation testing, subjected to solid–liquid separation, and vacuum-dried at a low temperature. The contact angle was measured using the sessile drop method, and the measured contact angle data were analyzed using ImageJ.js software.

2.6.3. Process Mineralogy Parameter Testing Using a Mineral Liberation Analyzer (MLA)

The raw ore was crushed and ground to a fineness of 70% passing 75 μm. Subsequently, an appropriate amount of the copper ore sample was ground into polished thin sections, which were analyzed using an MLA (Zhejiang Lante Optics Co., Ltd., Jiaxing, China) to determine the particle size of copper sulfides and pyrite and their liberation and intergrowth relationships.

2.6.4. XPS Measurement

The spectra of Fe and S elements on the surfaces of pyrite from different ore sites were measured using XPS (PHI 5000 VersaProbe III, ULVAC-PHI, Inc., Chigasaki, Japan) to determine the surface properties and elemental contents of the target minerals.

Full-spectrum scans of pyrite from different ore sites, narrow spectra of Fe and S on the pyrite surface, and raw XPS surface elemental data were obtained. Avantage 5.9931 BPM Foundation software was used to fit the spectral data and analyze the binding energies and contents of Fe and S on the surface.

3. Results and Discussion

3.1. Effects of Different Inhibition Systems on Pyrite Flotation

3.1.1. Effect of Lime on the Floatability of Pyrite in the Z200 System

Pyrite that had been ultrasonically cleaned and re-slurried was subjected to pH adjustment using lime. The dosages of Z200 and the frother methyl isobutyl carbinol (MIBC) were 10^{-3} mol/L and 10^{-4} mol/L, respectively. The dosages were determined based on the results of previous conditional tests. Figure 4 shows the test process and results.

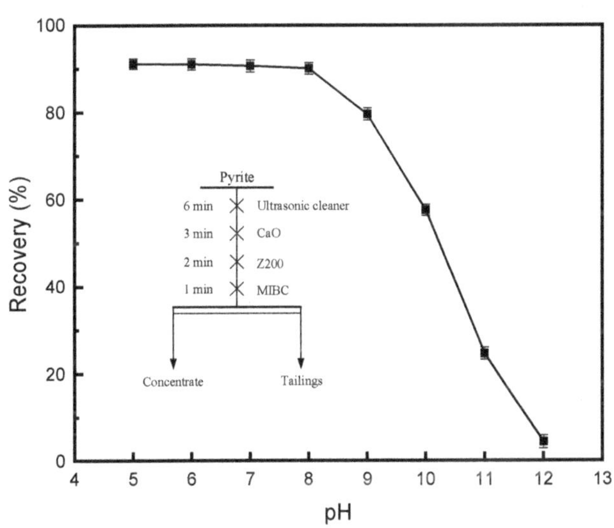

Figure 4. Effect of Lime on Pyrite Recovery.

As shown in Figure 4, lime exerted a strong inhibitory effect on pyrite flotation in the Z200 system. An increase in pH from 5.27 to 12 led to a gradual reduction in concentrate recovery, with pyrite recovery decreasing from 91.4% to 4.26%. However, it is evident that effective depression of pyrite requires a higher pH (>11) level to achieve.

3.1.2. Effect of H_2O_2 on Pyrite in the Z200 System

Given that pyrite is easily oxidized and H_2O_2 exerts a strong oxidizing effect, a hydrophilic film is readily formed on the surface of pyrite when reacted with H_2O_2, thus producing an inhibitory effect. During this test, natural slurry (pH = 4.37) that had not been pH-adjusted was treated with H_2O_2 as the inhibitor, Z200 as the collector (dosage: 10^{-3} mol/L), and MIBC as the frother (dosage: 10^{-4} mol/L). Figure 5 shows the test process and test results, respectively.

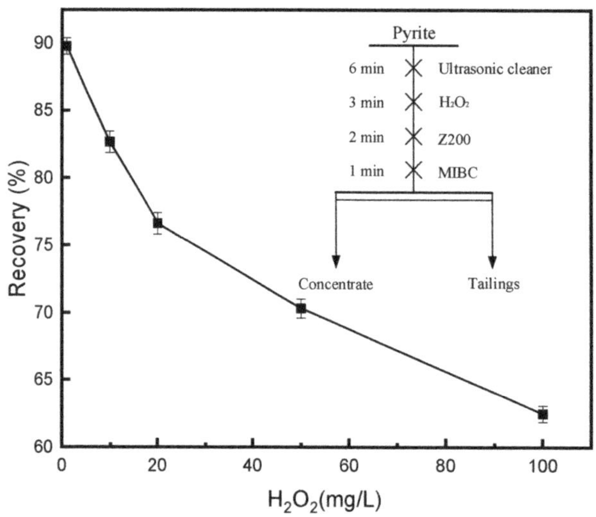

Figure 5. Effect of Hydrogen Peroxide Dosage on Pyrite Recovery.

As shown in Figure 5, the recovery rate decreased gradually from 89.79% to 62.51% when the H_2O_2 dosage was increased from 1 mg/L to 100 mg/L in the Z200 system. Recovery exceeded 80% at a H_2O_2 dosage of <10 mg/L but decreased to 62.5% when the H_2O_2 dosage was increased to 100 mg/L. These results indicated that the depression effect of standalone H_2O_2 on pyrite was not pronounced.

3.1.3. Effect of Combined Lime-H_2O_2 Inhibition on Pyrite in the Z200 System

To avoid excess alkalinity in the flotation solution system and excessive consumption of H_2O_2, lime + H_2O_2 was used as a combined inhibitor. Z200 was used as the collector (dosage: 10^{-3} mol/L), and MIBC was used as the frother (dosage: 10^{-4} mol/L). Figures 6 and 7 show the test process and test results, respectively.

Figure 7 shows that a combination of lime and H_2O_2 exerted an excellent inhibitory effect. The flotation recovery of pyrite decreased significantly with an increase in the pH of the solution environment. Pyrite recovery also exhibited a certain decline with increasing H_2O_2 dosage; However, excessively high pH and H_2O_2 dosage will inhibit covellite as Figure 7 shows. Therefore, pH = 10 and H_2O_2 dosage = 50 mg/L were ultimately selected as the optimal conditions, with a corresponding pyrite recovery of 24.67%.

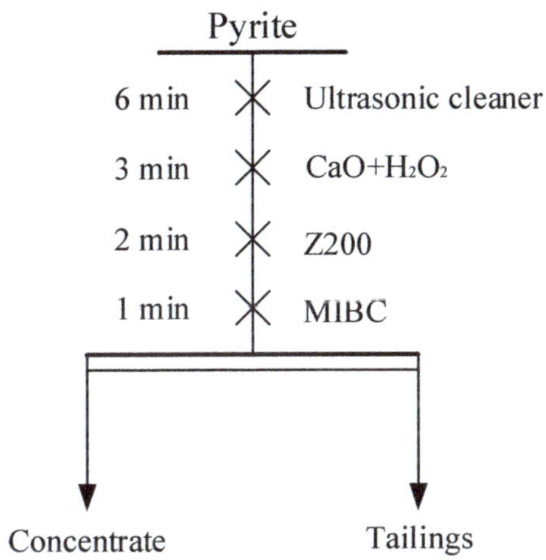

Figure 6. Flowchart of Single Mineral Flotation Test.

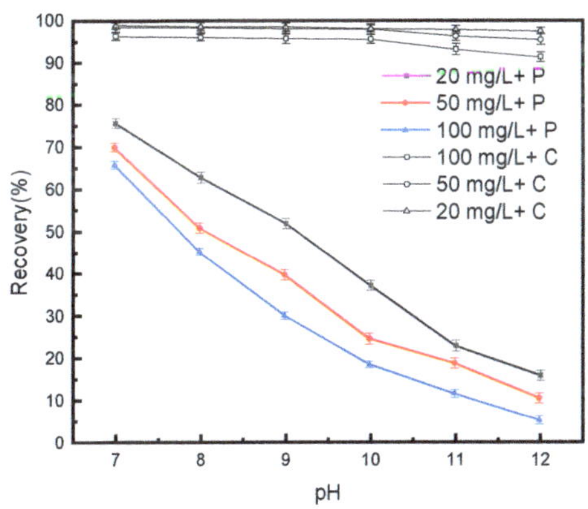

Figure 7. Effect of Lime (pH Adjustment) + Hydrogen Peroxide Mixed Depressant on Pyrite (P) and Covellite (C) Recovery.

3.2. Activation of Depressed Pyrite

To recover the inhibited pyrite, the effects of different activators on pyrite were investigated systematically. Sulfuric acid and copper sulfate were used to investigate the effect of activator dosage on the floatability of the inhibited pyrite.

3.2.1. Sulfuric Acid Activation Test

Using NaBX as the collector, pyrite recovery was evaluated with respect to the sulfuric acid dosage (pH). First, slurry pH was adjusted to 12 using the combined lime + H_2O_2 inhibitor, with the H_2O_2 dosage being 50 mg/L. Sulfuric acid was added for pH (from 10 to 2) adjustment after stirring. A butyl xanthate concentration of 1×10^{-4} mol/L was adopted, and MIBC was used as the frother. Figure 8 shows the test results.

As shown in Figure 8, in the system consisting of NaBX as the collector and lime + H_2O_2 as the combined inhibitor, a gradual increase in sulfuric acid dosage led to an initial increase and subsequent decrease in pyrite recovery. Optimal activation of the inhibited pyrite was achieved at a slurry pH of approximately 6, with recovery reaching 80%. Therefore, reducing the pulp pH can restore the floatability of pyrite.

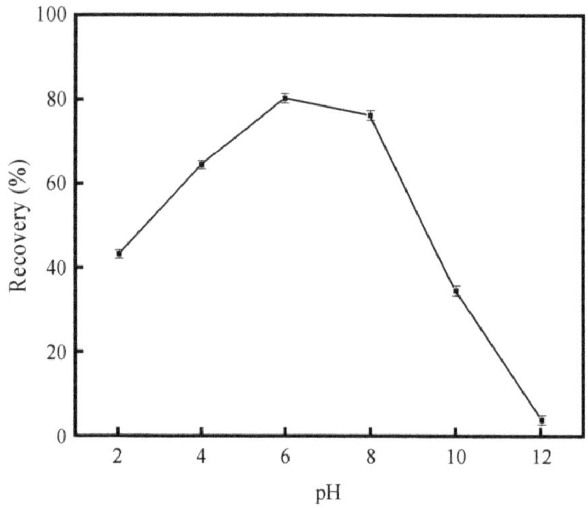

Figure 8. Effect of Sulfuric Acid (pH Adjustment) Dosage on Flotation Recovery of Inhibited Pyrite under NaBX System.

3.2.2. Copper Sulfate Activation Test

Using NaBX as the collector, the effect of copper sulfate dosage on pyrite recovery was investigated. First, slurry pH was adjusted to 10 using the combined lime + H_2O_2 inhibitor, with the H_2O_2 dosage being 50 mg/L. Copper sulfate is added at dosage from 1 to 10×10^{-4} mol/L. A NaBX concentration of 1×10^{-4} mol/L was adopted, and MIBC was used as the frother. Figure 9 shows the test results.

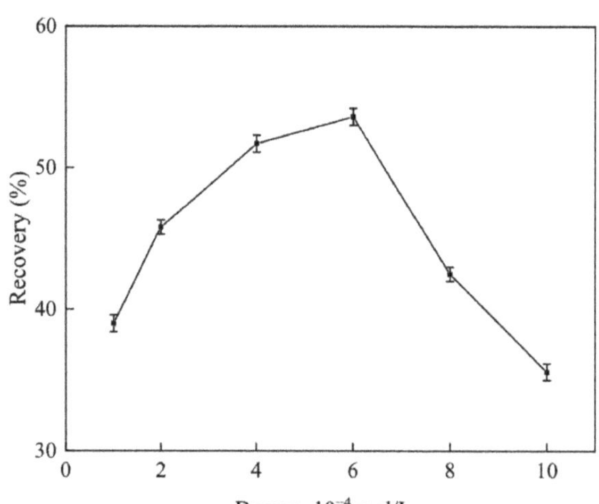

Figure 9. Effect of Copper Sulfate Dosage on the Floatability of Inhibited Pyrite under the Z200 System.

As shown in Figure 9, the recovery of pyrite inhibited via a combination of lime and H_2O_2 exhibited an initial increase and subsequent decrease with increases in the copper sulfate dosage when NaBX was used as the collector. This may be attributed to the large number of residual copper ions in the slurry. The consumption of NaBX by excessive copper sulfate led to a decrease in the NaBX concentration, resulting in a decline in collection effect and poorer activation by copper sulfate. A comparative analysis revealed that sulfuric acid provided a better activation effect than that of copper sulfate, because the recovery rate of pyrite obtained through sulfuric acid activation is much higher. Compared to the test results of sulfuric acid restoring pyrite floatability, the activation effect of copper sulfate is somewhat inferior; therefore, sulfuric acid is the more suitable reagent.

3.3. *XPS Analysis*

To further elucidate the intrinsic causes of pyrite inhibition by lime + H_2O_2, an XPS analysis was performed to evaluate the elemental components and chemical states of the surfaces of pure pyrite mineral treated with the combined inhibitor (lime + H_2O_2) and lime alone. Table 7 and Figure 10 show the analysis results.

Table 7. XPS Analysis Results of Pyrite before and after Depressant Treatment.

Species		B.E. (eV)	Atomic Concentration (at. %)		
			Blank	CaO	H_2O_2 + CaO
Fe $2p_{3/2}$	Fe(II)-S	707.2 ± 0.1	30.2	27.1	26.5
	Fe(III)-S	709.2 ± 0.3	5.7	4.3	4.1
	Fe(III)-O	710.3 ± 0.2	4.5	7.23	8.27
S $2p_{3/2}$	S_2^{2-}	162.9 ± 0.1	21.3	20.8	19.4
	S^{2-}	162.5 ± 0.2	10.5	9.2	9.6
	S_n^{2-}/S_0	164.3 ± 0.2	3.1	4.6	4.1
	SO_4^{2-}	168.7 ± 0.1	1.2	1.8	2.6
	SO_3^{2-}	166.5 ± 0.1	-	-	1.13
O 1s	O-H	532.8 ± 0.2	8.0	24.5	25.3
	Me-O	530.1 ± 0.2	2.3	12.8	13.6
Ca 2p	Ca-O	348.1 ± 0.2	-	0.1	0.1

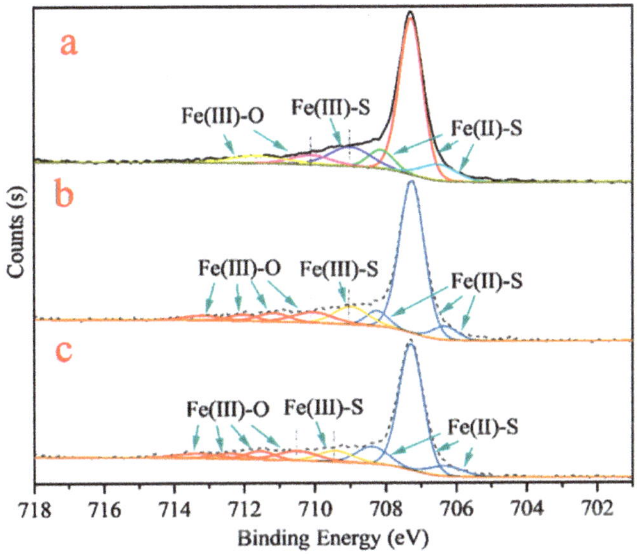

Figure 10. High-resolution Fe $2p_{3/2}$ spectra on the surface of pyrite under different conditions: (**a**) Blank, (**b**) Lime, and (**c**) Lime combined with hydrogen peroxide.

As shown in Table 7, the contents of O existing as O-H and Fe(III)-O on the surface of untreated pyrite were 8.0% and 4.5%, respectively. After lime treatment, the values increased considerably to 24.5% and 7.23%, respectively. Calcium ions were also detected on the pyrite surface. This suggested that the addition of lime caused the generation of a large number of hydroxides, which were co-adsorbed with calcium hydroxyl compounds to the pyrite surface and contributed to its hydrophilicity. These results are consistent with those reported by Duan Wenting et al. [27]. Upon treatment with the combined lime + H_2O_2 inhibitor, the surface atomic concentrations of pyrite underwent further change. First, the atomic concentrations of the sulfur element existing as S^{2-} and $S_2{}^{2-}$ decreased slightly, while those of sulfur existing as sulfates and sulfites increased. In addition, O contents in the form of O-H and Fe(III)-O increased by 1.04% and 0.8%, respectively, compared with those for lime treatment alone. These results indicated that oxidation occurred on the pyrite surface, leading to an increase in hydrophilic substances. This oxidation reaction was an inevitable outcome of H_2O_2 addition, with a greater hydrophilic effect on the pyrite surface induced by the combined lime + H_2O_2 inhibitor than that induced by lime alone.

Figure 10 shows the narrow spectrum of Fe $2p_{3/2}$ on the pyrite surface. The main peak at 707.2 ± 0.1 eV was attributed to fully coordinated low-spin Fe(II)-S from the bulk position, while the two small peaks at 706.3 ± 0.2 eV and 708.2 ± 0.2 eV were designated as multiple splitting peaks of Fe(II)-S [28]. The peak located at 709.3 ± 0.3 eV may be related to Fe(III)-S due to the simultaneous oxidation of Fe^{2+} to Fe^{3+} and reduction in S on the surface to S^{2-} (Equation (5)) [28–30]. Broad peaks with binding energies of 710.3 ± 0.2, 711.4 ± 0.2, 712.2 ± 0.2, and 713.2 ± 0.1 eV were attributed to the multiple splitting peaks of the oxygen or hydroxyl compounds of Fe^{3+} (i.e., FeOOH) [28,31–33].

Comparison of curves a and b in Figure 10 revealed that the Fe(III)-O peak on the pyrite surface was strengthened after lime treatment, indicating a significant increase in the hydroxyl oxides of iron. However, the binding energies of iron in various species remained almost unchanged compared with that of blank pyrite. These results suggested that the adsorption of lime on the surface of pyrite did not cause significant changes to the chemical environment of elemental iron, leading to a low degree of chemical reaction. However, different results were obtained when curves c and b were compared. The Fe(III)-O peak was further strengthened, and the inner electron binding energies of iron in both Fe(III)-O and Fe(III)-S were clearly shifted to the left and right. This suggested that significant changes occurred in the chemical environment of iron. Such a phenomenon can be attributed to the oxidative effect of H_2O_2 on the pyrite surface, with surface oxidation resulting in the conversion of the main surface iron species Fe(II)-S to Fe(III)-O (Equations (2)–(4)) [34].

$$Fe^{2+}_{surface} + S^{-}_{surface} \rightarrow Fe^{3+}_{surface} + S^{2-}_{surface} \quad (2)$$

$$FeS_2 + \frac{15}{4}O_2 + \frac{7}{2}H_2O \rightarrow Fe(OH)_3 + 2SO_4^{2-} + 4H^+ \quad (3)$$

$$FeS_2 + \frac{15}{4}O_2 + \frac{5}{2}H_2O \rightarrow FeOOH + 2SO_4^{2-} + 4H^+ \quad (4)$$

Figure 11 shows the S 2p spectra of the pyrite surface, with binding energies of 162.5 ± 0.3, 162.8 ± 0.1, and 164.4 ± 0.2 eV for S 2p3/2 corresponding to monosulfides (S^{2-}), disulfides ($S_2{}^{2-}$), and hydrophobic polysulfides ($S_n{}^{2-}/S_0$), respectively [35,36]. Sulfur oxide species, such as sulfites ($SO_3{}^{2-}$) and sulfates ($SO_4{}^{2-}$), exhibited S $2p_{3/2}$ peaks at 166.55 and 168.7 ± 0.1 eV, respectively [35,37]. In untreated pyrite, the concentrations of $SO_3{}^{2-}$ and $SO_4{}^{2-}$ were 0% and 1.2%, respectively. After lime treatment, the $SO_4{}^{2-}$ concentration increased slightly from 1.2% to 1.8%. Treatment with the combined lime + H_2O_2 inhibitor led to a further increase in the pyrite surface $SO_4{}^{2-}$ concentration to 2.6%. In addition to the occurrence of new $SO_3{}^{2-}$ species, the S 2p peak in $S_2{}^{2-}S_n{}^{2-}/S_0$

was shifted leftward. These results suggested further enhancement of the pyrite surface oxidation process (Equations (5) and (6)).

$$S_2^{2-} - e^-(\text{to } Fe^{3+}) + H_2O \rightarrow SO_3^{2-} + Fe^{2+} + H^+ \quad (5)$$

$$2SO_3^{2-} + O_2 + 2H^+ \rightarrow 2SO_4^{2-} + H_2O \quad (6)$$

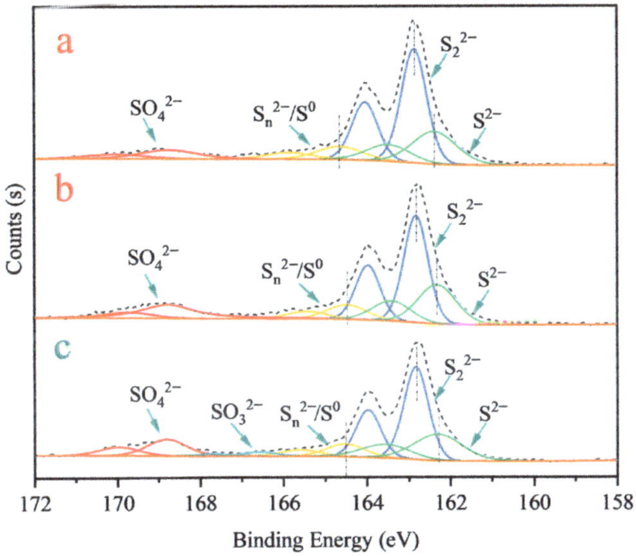

Figure 11. Narrow Scan of S 2p on the Pyrite Surface under different conditions: (**a**) Blank, (**b**) Lime, and (**c**) Lime combined with hydrogen peroxide.

Therefore, the XPS analysis results indicated that the addition of H_2O_2 increased the concentration of hydrophilic iron oxide/hydroxide species on the pyrite surface.

3.4. SEM Analysis

Using SEM, the microscopic morphology of mineral surfaces was observed directly, providing a basis for determining the effects of different inhibitor conditions on the pyrite surface.

From the surface morphologies of pyrite under different conditions shown in Figure 12, it can be seen that the raw pyrite surface was relatively uniform. A certain amount of small debris was attached to the surface, and there were no obvious traces of oxidation or other chemical reactions. Figure 12b shows the spectrum of pyrite that had been reacted with an aqueous lime solution at a certain concentration. Compared with the raw ore, the pyrite surface remained nearly unchanged except for the presence of two natural small holes. Figure 12c shows the surface morphology of pyrite after reaction with H_2O_2. A large number of corrosion pits appeared on the pyrite surface. This indicated the occurrence of a strong chemical reaction between the pyrite surface and H_2O_2, resulting in oxidation of the mineral surface. Figure 12d shows the surface morphology of pyrite after reaction with lime + H_2O_2. In addition to the presence of corrosion pits on the surface, small bubble-like substances were also observed, likely to be caused by hydroxide deposition on the surface. The above results demonstrated that the addition of H_2O_2 led to oxidation of the pyrite surface, and the combined use of H_2O_2 with lime also significantly influenced the surface morphology of the mineral.

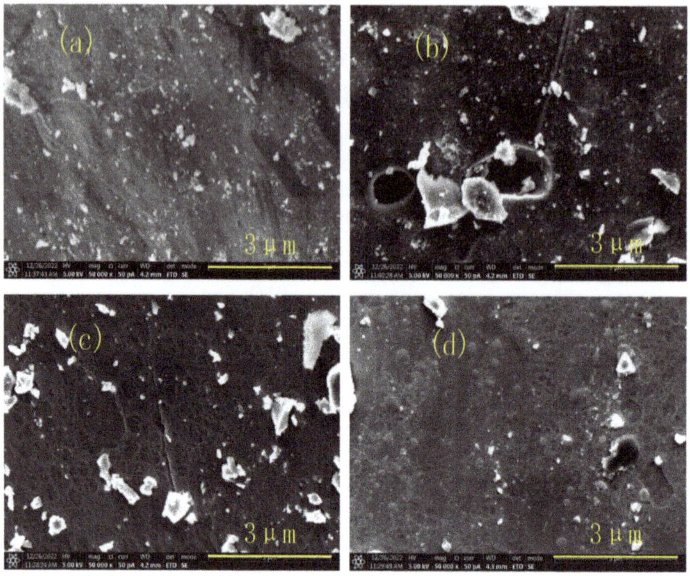

Figure 12. Scanning Electron Microscope Images of Pyrite Samples: (**a**) Pyrite; (**b**) Pyrite + Lime Water; (**c**) Pyrite + Hydrogen Peroxide; (**d**) Pyrite + Hydrogen Peroxide + Lime Water.

3.5. Contact Angle Analysis

The effects of lime + H_2O_2 on pyrite floatability were investigated based on analyses of the contact angle of the pure pyrite mineral surface before and after treatment with the combined inhibitor. Table 8 shows the results of contact angle measurement and analysis.

Table 8. Changes in the Contact Angle of Pyrite Before and After Depressant Application.

Test Condition	Pyrite	
	Appearance of Contact Angle	**Contact Angle (°)**
Fresh Surface	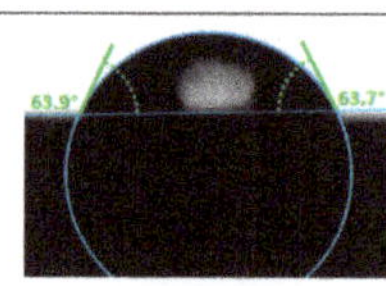	63.8
Lime		48.3
Lime + Hydrogen Peroxide		40.5

The contact angle of the surface of untreated pyrite was 63.8°, indicating good floatability. After lime treatment, the contact angle was reduced to 48.3°, indicating the enhancement of hydrophilicity. Treatment with the combined lime + H_2O_2 inhibitor led to a significant decrease in surface contact angle to 40.5°, indicating a considerable increase in hydrophilicity. When viewed in combination with the XPS analysis results, these findings demonstrated that the increase in the contents of oxides and hydroxides generated on the surface of pyrite resulted in increased hydrophilicity.

3.6. Validation Through Actual Ore Flotation Testing

Based on the optimal parameters for pyrite inhibition and activation determined from single-mineral laboratory flotation testing, copper ore from a mine in Serbia was used as the raw material, Z200 was used as the copper collector, lime + H_2O_2 was used as the combined inhibitor, sulfuric acid was used as the activator, and NaBX was used as the sulfur collector for validation of the feasibility of the process protocol described above for the comprehensive separation and recovery of pyrite and covellite. Figure 1 shows the test process.

The flotation test results shown in Table 9 indicated that a change in the pyrite inhibitor from lime alone to lime + H_2O_2 led to an increase in the copper concentrate grade from 30.35% to 31.12%, with recovery remaining almost unchanged. The recovery of pyrite increased from 56.70% to 58.99%, while that of gold increased by nearly 4%. Therefore, the use of the combined lime + H_2O_2 inhibitor effectively increased gold recovery with remarkable results.

Table 9. Flotation Test Results of a Copper–Gold Mine in Serbia.

Product	Yield (%)	Grade (%)			Recovery (%)		
		Cu	Au ($g \cdot t^{-1}$)	S	Cu	Au	S
Copper Concentrate 1	7.42	31.12	4.13	34.6	80.74	18.69	12.59
Copper Concentrate 2	4.68	8.51	4.58	36.8	13.93	13.07	8.45
Pyrite Concentrate	24.35	0.47	3.92	49.4	4.00	58.20	58.99
Tailing 2	14.53	0.084	0.5	7.61	0.43	4.43	5.42
Tailing 1	49.02	0.05	0.19	6.05	0.91	5.61	14.55
Feed	100	2.86	1.64	20.39	100	100	100
Copper Concentrate 1 + 2	12.33	22.37	4.30	/	94.66	31.76	21.04
Tailing 1 + 2	63.55	0.06	0.26	6.41	1.34	10.04	19.97

4. Conclusions

1. Micro-flotation test results indicate that the addition of H_2O_2 reduces the pH required for pyrite suppression from 12 to 10. The combined use of lime and H_2O_2 as an inhibitor system enables efficient separation of covellite from pyrite.
2. An investigation into the mechanisms revealed that the added H_2O_2 oxidized and corroded the pyrite surface, leading to increased oxide and hydroxide generation. This contributed to enhanced surface hydrophilicity and thereby to effective pyrite inhibition.
3. Pyrite that has been inhibited can partially regain its floatability upon the addition of sulfuric acid or copper sulfate, with sulfuric acid demonstrating greater effectiveness.
4. The Serbian copper ore exhibits a high sulfur content. When lime is used alone for pyrite inhibition, pyrite and gold recovery rates remain low. However, employing lime and H_2O_2 as a combined inhibitor reduces lime consumption in industrial operations, maintains copper flotation performance, and increases pyrite recovery by 2.3% and gold recovery by nearly 4%.

Author Contributions: Conceptualization, Y.Y., M.K. and R.D.; methodology, M.K., Y.Y. and R.D.; formal analysis, Y.Y., Y.L. and R.D.; resources, Y.Y. and M.K.; data curation, Y.Y.; writing—original draft preparation, Y.Y.; writing—review and editing, M.K., R.D. and Y.L.; visualization, Y.L.; supervision, M.K.; project administration, M.K. and R.D. All authors have read and agreed to the published version of the manuscript.

Funding: This research received no external funding.

Data Availability Statement: The data obtained and analyzed within the frame of this study will be made available by the corresponding author upon reasonable request.

Conflicts of Interest: Authors Yuankun Yang and Yinying Liao were employed by the company Zijin Mining Group Co., Ltd. The remaining authors declare that the research was conducted in the absence of any commercial or financial relationships that could be construed as a potential conflict of interest.

Abbreviations

The following abbreviations are used in this manuscript:

MLA	Mineral liberation analyzer
SEM	Scanning electron microscopy
XPS	X-ray photoelectron spectroscopy

References

1. Luo, S.X.; Chen, H.; Mou, Q.S.; Wu, Y.H. Research situation and progress of adsorption properties of pyrite. *Multipurp. Util. Miner. Resour.* **2020**, *5*, 27–33.
2. Chen, W.; Chen, T.; Bu, X.; Chen, F.; Ding, Y.; Zhang, C.; Deng, S.; Song, Y. Theselective flotation of chalcopyrite against galena using a lginatesas depressant. *Miner. Eng.* **2019**, *141*, 105848. [CrossRef]
3. Wang, F.L. Basic Theoretical Research on Flotation of Main Sulfide Oxide Minerals of Copper, Lead, Zinc and Iro. Ph.D. Thesis, Northeastern University, Shenyang, China, 2008.
4. Chimonyo, W.; Fletcher, B.; Peng, Y. The differential depression of anoxidized starch on the flotation of chalcopyrite and graphite. *Miner. Eng.* **2020**, *146*, 106114. [CrossRef]
5. Schlesinger, M.E.; Sole, K.C.; Davenport, W.G.; King, M.J. Chapter 1—Overview. In *Extractive Metallurgy of Copper*, 5th ed.; Schlesinger, M.E., King, M.J., Sole, K.C., Davenport, W.G., Eds.; Elsevier: Amsterdam, The Netherlands, 2011; pp. 1–12.
6. Wang, J.; Zheng, S.; Liu, W.; Chen, L.; Wen, Z.; Li, X. Prediction, evaluation and optimization of China's copper resource supply system under carbon constraints. *Sustain. Prod. Consum.* **2023**, *39*, 285–300. [CrossRef]
7. Li, W.Y.; Bai, X.; Ge, F.; Quan, X.; Chen, M.J.; Hao, J.M. Development Status of Copper-sulfur Separation Influence Factors and Depressants. *Multipurp. Util. Miner. Resour.* **2025**, 1–18.
8. Long, Y.Y.; Yin, W.Z.; Long, K.Y. Synergistic oxidation of NaClO and $KMnO_4$ on flotation separation of chalcopyrite and pyrite and mechanism therein. *Min. Metall. Eng.* **2024**, *44*, 51–56.
9. Tang, X.; Chen, Y. A review of flotation and selective separation of pyrrhotite: A perspective from crystal structures international. *J. Min. Sci. Technol.* **2022**, *32*, 847–863. [CrossRef]
10. Shan, Z.Q.; Xia, Y.; Zhang, X.H. Copper selection experimental on a copper-gold sulphide Ore in Balai County, Lao People's Democratic republic. *Met. Mine* **2016**, *9*, 96–99.
11. Chen, J.H.; Li, Y.Q.; Long, Q.R. Molecular structures and activity of organic depressants for marmatite, jamesonite and pyrite flotation. *Trans. Nonferrous Met. Soc. China* **2010**, *20*, 1993–1999. [CrossRef]
12. Bi, Y.X.; Yu, P.; Ding, Z.; Bai, S.J.; Wen, S.M. The Development of research on the pyrite flotation depressants. *Conserv. Util. Miner. Resour.* **2020**, *40*, 157–166.
13. Cao, Q.; Zhang, H.; Yan, Y.; Li, Y.; Liu, D. Flotation separation of pyrite and chalcopyrite with potassium permanganate as a depressant. *Chem. Pap.* **2024**, *78*, 1761–1773. [CrossRef]
14. Mu, Y.; Peng, Y.; Lauten, R.A. The depression of pyrite in selective flotation by different reagent systems—A Literature review. *Miner. Eng.* **2016**, *96–97*, 143–156. [CrossRef]
15. Li, S.H.; Ma, Q.; Wang, L.; Yang, X. Research Progress on Activation Mechanism of Pyrite Flotation. *Multipurp. Util. Miner. Resour.* **2023**, *2*, 124–130.
16. Qi, X.; Li, X.; Liang, Y.; Wang, H.; Guo, W.; Cong, X.; Lv, F.; Zhang, H. Surface structure-dependent hydrophobicity/oleophilicity of pyrite and its influence on coal flotation. *J. Ind. Eng. Chem.* **2020**, *87*, 136–144. [CrossRef]
17. Bi, Y. Research on Selective Flotation Separation of Chalcopyrite and Pyrite Under Low Alkalinity. Ph.D. Thesis, Kunming University of Science and Technology, Kunming, China, 2023.

18. Khoso, S.A.; Fei, L.Ü.; Ya, G.A.O.; Wei, S.U.N. Xanthate interaction and flotation separation of H_2O_2-treated chalcopyrite and pyrite. *Trans. Nonferrous Met. Soc. China* **2019**, *29*, 2604–2614. [CrossRef]
19. Wang, Y.; Wei, D.; Wen, S.; Lin, Q.; Song, Z.; Luo, X. Enhanced pyrite depression through nitrogen-aerated flotation under low alkalinity conditions. *Adv. Powder Technol.* **2024**, *35*, 104424. [CrossRef]
20. Ozun, S.; Hassas, B.V.; Miller, J.D. Collectorless flotation of oxidized pyrite. *Colloids Surf. A Physicochem. Eng. Asp.* **2019**, *561*, 349–356. [CrossRef]
21. Luo, X.; Yang, W.; Wen, S.; Lin, Q.; Wei, D.; Li, C.; Zhang, Y.; Song, Z.; Wang, Y. Effect of regulating dissolved oxygen concentration in pulp with aerated gas on pyrite flotation. *Miner. Eng.* **2023**, *202*, 108290. [CrossRef]
22. Feng, B.; Zhang, W.; Guo, Y.; Wang, T.; Luo, G.; Wang, H.; He, G. The flotation separation of galena and pyrite using serpentine as depressant. *Powder Technol.* **2019**, *342*, 486–490. [CrossRef]
23. Lee, R.L.J.; Chen, X.; Peng, Y. Flotation performance of chalcopyrite in the presence of an elevated pyrite proportion. *Miner. Eng.* **2022**, *177*, 107387. [CrossRef]
24. Hodgkinson, G.; Sandenberght, R.F.; Hunter, C.J.; De Wet, J.R. Pyrite flotation from gold leach residues. *Miner. Eng.* **1994**, *7*, 691–698. [CrossRef]
25. Yuan, J.; Li, Y.; Ding, Z.; Zhang, Y.; Chen, L.; Wen, S.; Bai, S. Cleaner flotation separation of chalcopyrite from pyrite at low alkalinity: Based on calcium hypochlorite oxidative modification in acid mine drainage system. *Sep. Purif. Technol.* **2024**, *344*, 127325. [CrossRef]
26. Yang, X.; Li, Y.; Fan, R.; Duan, W.; Huang, L.; Xiao, Q. Separation mechanism of chalcopyrite and pyrite due to H_2O_2 treatment in low-alkaline seawater flotation system. *Miner. Eng.* **2022**, *176*, 107356. [CrossRef]
27. Duan, W.; Wang, Y.; Rao, F.; Liu, W.; Yang, Y.; Deng, R. Correlation between the mineralogical properties and oxidation rate of different minerogenetic pyrites: A Reflection on Floatability. *Min. Metall. Explor.* **2023**, *40*, 1369–1381. [CrossRef]
28. Schippers, A.; Jozsa, P.; Sand, W. Sulfur chemistry in bacterial leaching of pyrite. *Appl. Environ. Microbiol.* **1996**, *62*, 3424–3431. [CrossRef] [PubMed]
29. Holmes, P.R.; Crundwell, F.K. The kinetics of the oxidation of pyrite by ferric ions and dissolved oxygen: An electrochemical study. *Geochim. Cosmochim. Acta* **2000**, *64*, 263–274. [CrossRef]
30. Ejtemaei, M.; Nguyen, A.V. Characterisation of sphalerite and pyrite surfaces activated by copper sulphate. *Miner. Eng.* **2017**, *100*, 223–232. [CrossRef]
31. Nesbitt, H.W.; Bancroft, G.M.; Pratt, A.R.; Scaini, M.J. Sulfur and iron surface states on fractured pyrite surfaces. *Am. Mineral. Gruyter* **1998**, *83*, 1067–1076. [CrossRef]
32. Niu, X.; Chen, J.; Li, Y.; Xia, L.; Li, L.; Sun, H.; Ruan, R. Correlation of surface oxidation with xanthate adsorption and pyrite flotation. *Appl. Surf. Sci.* **2019**, *495*, 143411. [CrossRef]
33. Nesbitt, H.W.; Muir, I.J. X-ray photoelectron spectroscopic study of a pristine pyrite surface reacted with water vapour and air. *Geochim. Cosmochim. Acta* **1994**, *58*, 4667–4679. [CrossRef]
34. Nesbitt, H.W.; Muir, I.J.; Prarr, A.R. Oxidation of arsenopyrite by air and air-saturated, distilled water, and implications for mechanism of oxidation. *Geochim. Cosmochim. Acta* **1995**, *59*, 1773–1786. [CrossRef]
35. Bonnissel-Gissinger, P.; Alnot, M.; Ehrhardt, J.J.; Behra, P. Surface oxidation of pyrite as a function of pH. *Environ. Sci. Technol.* **1998**, *32*, 2839–2845. [CrossRef]
36. Chiriță, P.; Schlegel, M.L. Pyrite oxidation in air-equilibrated solutions: An electrochemical study. *Chem. Geol.* **2017**, *470*, 67–74. [CrossRef]
37. Cai, Y.; Pan, Y.; Xue, J.; Sun, Q.; Su, G.; Li, X. Comparative XPS study between experimentally and naturally weathered pyrites. *Appl. Surf. Sci.* **2009**, *255*, 8750–8760. [CrossRef]

MDPI AG
Grosspeteranlage 5
4052 Basel
Switzerland
Tel.: +41 61 683 77 34

Minerals Editorial Office
E-mail: minerals@mdpi.com
www.mdpi.com/journal/minerals

www.ingramcontent.com/pod-product-compliance
Lightning Source LLC
Chambersburg PA
CBHW061601210926
51850CB00009B/2180

* 9 7 8 3 7 2 5 8 7 6 1 5 0 *